THE WORLD OF
Northern Evergreens

THE WORLD OF
Northern Evergreens

E. C. Pielou

Comstock Publishing Associates
a division of Cornell University Press
ITHACA AND LONDON

Library of Congress Cataloging-in-Publication Data

Pielou, E. C., 1924–
    The world of northern evergreens.
    Bibliography: p.
    Includes index.
    1. Conifers—North America. 2. Evergreens—North
America. 3. Forest ecology—North America. I. Title.
QK494.P54 1988   585′.2′097   87-47966
ISBN 0-8014-2116-0
ISBN 0-8014-9429-9 (pbk.)

I thank Cynthia Page for modifying Figures 112–115,
117–119, 121–123, and 126–129.

First published 1988 by Cornell University Press.
Comstock|Cornell Paperbacks edition first published 1988.
Printed in the United States of America

*The paper in this book is acid-free and meets the guidelines for
permanence and durability of the Committee on Production Guidelines
for Book Longevity of the Council on Library Resources.*

For Chelsea and Kayley

Contents

Preface

For many people, certainly for the majority of North Americans with homes in the northern half of the continent, coniferous trees constitute a large fraction of all the "living material" they will see in a lifetime. Naturalists, hikers and backpackers, canoeists and cross-country skiers, fishermen and hunters—in fact, all whose work or recreation takes them outdoors in northern North America—are accustomed to seeing coniferous trees by the tens of millions, whether they consciously notice them or not.

Outdoor people have a wide spectrum of interests. There are many kinds of naturalists—for example, birders, butterfly collectors, rock hounds, and plant-hunters—and within each group are specialists and superspecialists. Specialists tend to specialize in "interesting" items: a birder is more likely to concentrate on owls, for instance, than on starlings; most butterfly collectors are bored with cabbage whites; and the average plant-hunter finds orchids more fascinating than crabgrass. Because of this preoccupation with the hard-to-find, the beautiful, and the unusual, most of the commonest objects in nature are apt to be ignored. They are simply *there*, part of the background. But to assume that because a thing is common it is therefore uninteresting is a mistake.

For most people, familiarity breeds indifference (contempt is too strong a word), and for most outdoor people the fact that they will no doubt encounter rank upon rank of coniferous trees in excursion after excursion in the future neither pleases nor displeases them. They don't even think about it. If there are innumerable coniferous trees in your future (and this is surely true for all northern naturalists), why not take advantage of the fact, look at the trees more closely, and learn something about them? Knowledge cannot fail to bring interest and appreciation.

This book is designed to draw the attention of northern naturalists to some of the many interesting things to be observed on any journey that takes them among evergreen trees and larches (the only deciduous conifers in the North). For the purpose of the book, "northern North America" means all of the continent north of 45° north latitude. All the conifers to be found growing wild in at least some part of this enormous region are mentioned in the following pages. This does not mean, of course, that they do not also occur south of the 45th parallel. Many of the trees in our region have ranges extending a long distance southward, especially in the mountains of the West. Some even occur as far south as Mexico. More is said about their ranges at the end of Chapter 3.

Learning to identify the different species of coniferous trees is only a beginning. Once you know the trees, many things can be observed if you know what to look for. There is, however, a world of difference between seeing and interpreting. Even though everyone *sees* the same things, it doesn't follow that everyone *interprets* them correctly or understands them. The ability to interpret is the hallmark of the true naturalist, and developing that ability is one of the pleasures of being a naturalist. The well-informed naturalist understands and enjoys a thousand things that the uninformed one doesn't even notice; and the more people who understand and enjoy the woods, the more there will be to protect them.

E. C. PIELOU

Denman Island, British Columbia

Chapter 1

Origins of the Evergreen Forests

Of all the people who enter an evergreen forest, only a handful ever ask themselves these two questions: Where has the forest come from? And why is it composed of evergreens rather than deciduous, broad-leaved, hardwood trees?

The answers are by no means obvious. Indeed, the questions have puzzled ecologists and motivated years of careful investigation. The results of this research provide at least partial answers; and although the pieces of evidence on which the answers depend are not, for the most part, observable on a hike in the woods, the hike would certainly be more interesting for somebody knowing a bit about the questions, even if only that they are questions.

CONIFERS AND THE ICE AGE

Consider the first of the two questions: Where has the forest come from? The only certain answer is that the trees in nearly all of our area must have descended from ancestors that lived a long distance away.

About 18,000 years ago, when the ice sheets of the most recent ice age had reached their maximum extent, they covered nearly all of northern North America. The shaded area in Figure 1 shows the regions that were then under ice. The glaciated areas must have been like present-day Greenland and Antarctica—barren, unbroken expanses of ice, utterly devoid of plant life. Only a small fraction of our area escaped glaciation: parts of Yukon and Alaska in the Far Northwest, and a narrow marginal strip in the

Figure 1. Ice-covered area of North America during the last ice age.

Southwest. But conditions near the margin of the ice in the unglaciated regions must have been bleak. It therefore seems very likely (nothing is ever certain in reconstructions of the past) that, for a time, not a single evergreen tree was to be found anywhere in our area. The contrast with present-day is striking.

The climate began to warm up, and the ice to melt, about 18,000 years ago. As the ice sheets dwindled, the newly exposed land became available for plants and gradually, very gradually, evergreen forests invaded the area. Seeds were blown in from the south. For at the time of maximum glaciation, evergreen forests stretched across the continent south of the ice margin; even the Great Plains were forested. East of the continental divide the most abundant trees were spruces and Jack Pines (the evidence that allows us to visualize the forests of the distant past is described in Chapter 4). The forests south of the ice on the West Coast had a richer mix of tree species and were probably much like they are now.

The northward march of the forests into the newly ice-free land was inevitably slow. There was no soil to start with—only lifeless mixtures of boulders, gravel, sand, and clay, laid bare by the melting ice. The development of a soil adequate for trees must have taken considerable time. Different kinds of trees arrived to occupy their present geographic ranges at different times. Those that survived the ice age farther south than the spruces and Jack Pines had farther to come; and those that require a deep organic soil if they are to thrive arrived long after the species that can make do with thin, rocky soils. For example, hemlock is believed to have reached what is now northern Michigan about 3000 years later than white pine.[1]

The climate is still changing and trees are still migrating. The geographic range of every species changes continuously, as the trees invade new territory when it becomes hospitable, and die out in previously held territory when it becomes inhospitable. The climatic warming trend that brought an end to the ice age continued for about 10,000 years; it reached its maximum about 8000 years ago, even before the ice sheets had completely melted. Because of the tremendous length of time required for the melting of huge masses of ice, two remnants of the original vast ice sheets persisted, one to the east and one to the west of Hudson Bay, until about 6500 years ago. They finally disappeared *after* the time of maximum warmth. Since that maximum there has been a cooling trend; though it has not been uninterrupted, many climatologists see it as a definite trend and the onset of the next ice age as having already begun. Many of our evergreen tree species have already lost some ground; they do not, nowadays, grow as far north as they did at the time of maximum warmth.

The Advantage of Being Evergreen

The second of the two questions posed at the beginning of the chapter can be reworded as follows: Why is so much of the area that was ice-covered during the last ice age now covered with evergreen, or coniferous, forest? Why conifers (cone-bearing trees) rather than hardwoods? To say that conifers are better adapted to the environment is, of course, not an answer but merely a way of evading the question or, rather, of prompting another: In what way are they better adapted? The answer cannot be simple, because conifers do better than hardwoods in a number of quite different environments. They obviously thrive in the mild, moist climate of the Pacific coast; and they also prosper, if less luxuriantly, in the cold, harsh, rather dry climate of subarctic Canada.

Consider the West Coast forests first. The mild winters and abundant rain create an ideal environment for many trees, not only for conifers. Poplars, aspens, and alders (about which more in Chapter 10) flourish in coastal conditions, and grow much larger than do inland representatives of their species. But the magnificent conifers, the enormous trees that arouse the wonder of everyone who sees them, greatly outnumber the hardwoods. They have done better than the hardwoods in the struggle for existence, and have taken possession of a much larger share of the land.

It is believed that it is their evergreenness that makes conifers superior to hardwoods in the climate of the Pacific Northwest. Because they have green leaves all through the year, conifers can carry on photosynthesis whenever the weather is warm enough. They can profit from warm spells in late fall and early spring and even, in some years, in midwinter. The deciduous hardwoods, being leafless for half the year, cannot cash in on these fleeting opportunities. The period in which the temperature is warm enough for photosynthesis is longer than the period in which the deciduous hardwoods are in leaf.

For the trees of the West Coast, the most stressful time of the year comes during the summer drought. This is an unusual state of affairs, as everywhere else in our area winter is more stressful than summer. Water shortage affects both conifers and hardwoods, but the hardwoods are much less able to cope with this adversity because their big, thin leaves dry out much more easily than do the small, needlelike leaves of the conifers (more on this in Chapter 5). Thus the annual drought comes at a time when the deciduous hardwoods are most vulnerable, the time when they are in leaf.

Thus, though it seems clear why conifers do better than hardwoods in a climate with mild winters and dry summers, these arguments obviously don't hold in subarctic Canada, where conifers also predominate. The ability to carry on photosynthesis on any warm day whatever the season is not particularly useful to a tree growing where the only warm days, with rare exceptions, are in summer. There must be some other explanation for the success of conifers in the vastly different conditions of the Far North.

THE ADVANTAGE OF LONG-LIVED LEAVES

The success of evergreen conifers vis-à-vis hardwoods in the North probably has little to do with the harsh northern climate. The most likely reason is that conifers are better able to survive on inferior soils. In much of the area once covered by the ice sheets the soil is apt to be poor. Not enough time has elapsed since the end of the last ice age for deep, rich soils to develop, and the ground cover left when the ice melted has not so far acquired much organic material. Some soils are little better than dry sand, and others are merely thin, dusty coatings over hard bedrock.

The question that now arises is: Why should evergreens be better adapted than hardwoods to life on poor soils? The answer seems to be, once again, that evergreen leaves confer an advantage. In this case the advantage is that evergreen leaves last for several years. Therefore an evergreen conifer, unlike a deciduous tree, does not have to grow a whole new set of leaves every spring and its demands for nourishment are correspondingly less. The one deciduous conifer in our area, larch, is an awkward exception to these generalizations. Larch is believed to have descended from an evergreen ancestor several million years ago; but why a change from evergreenness to deciduousness should have been a benefit in the struggle for existence is not at all clear.

Compared with hardwoods, evergreen conifers lead far more "frugal" lives. They are not so efficient at photosynthesis, but then they don't need to be, since they don't have to synthesize the material for a new set of leaves each spring. Conifers retain a large proportion of the mineral nutrients they absorb from the soil and are therefore able to get by on comparatively infertile soils. Hardwoods, on the other hand, lose much of each summer's nutrient intake when they shed their leaves in the fall and they therefore require a richer soil to supply the necessary replacements.

To put the matter concisely, conifers live on a "waste-not–want-not" system, hardwoods on an "easy-come–easy-go" system. More technically, conifers cycle materials slowly, hardwoods comparatively fast. It is easy to see that materials-cycling proceeds rather slowly in, for example, a pine forest; a thick carpet of dry pine needles underfoot shows that the trees' discarded materials are slow to decay. Evergreen leaves, because of their tough texture, do not decay nearly as fast as the soft leaves of deciduous hardwoods. Therefore the nutrients within the leaves take longer to return to the soil in a conifer forest than they do in a hardwood forest. Indeed, the infertility of conifer forest soils is caused to some extent by the conifers themselves.

ENDURING THE COLD

If there is one adaptation that is more vitally important than any other to a northern tree, it is cold hardiness. Trees and tall shrubs have to endure the lowest air temperatures that winter brings, temperatures that would kill a lightly clothed person in a very short time. Herbaceous plants that die back to ground level in winter are protected by the soil they are imbedded in and insulated by the snow blanket that covers them. In cold weather their environment is much warmer than that of the trees and shrubs, which must be properly adapted to the cold if they are to survive.

All the trees that grow in our area, hardwoods as well as conifers, are hardy in the ordinary, gardener's sense of the word. That is, they are unharmed by comparatively mild frosts, provided they have had time to become dormant before freezing weather sets in. However, from a tree's point of view, there are two different intensities of frost: mild and severe, the dividing line being at about −40° Celsius. (As −40° is the same on both the Celsius and the Fahrenheit scales, we could as well say that the dividing line comes at −40° Fahrenheit.)

Our trees are adapted to escape frost injury in one or other of two entirely different ways: Some do it by *supercooling*, the remainder by *extracellular freezing*. Supercooling protects only at temperatures above −40°, whereas extracellular freezing is effective at any naturally occurring temperature, however low. Therefore, we have two kinds of trees, *hardy* and *very hardy*. The hardy trees are those that rely on supercooling to survive the winter, and consequently they cannot grow where minimum winter temperatures fall below −40°. The very hardy trees are those capable of extracellular freezing.

Now for the ways in which these frost-proofing adaptations work. In hardy trees (but not in very hardy ones) the liquids in the living cells become supercooled as the temperature drops. That is, the liquids remain liquid even below their customary freezing temperatures because there are no minute particles inside the cells, or rougherings on the inner faces of the cell walls, to act as nuclei around which ice crystals can begin to form. (The reason why ice won't crystallize, except at very low temperatures, when there is nothing for the first ice crystals to attach themselves to is a problem in the realm of physics; for example, snowflakes need particles of dust as nuclei around which to grow). But if the temperature drops below −40°, the cell liquids freeze anyway, with or without particles to act as nuclei, and the cells are killed.

The very hardy trees have a completely different way of surviving the cold. There are great numbers of small, empty spaces within the living tissues of a tree. Liquids inside the cells ooze out through the walls of the cells and freeze in these spaces, a process described as extracellular freezing. As the ice crystals are not inside the living cells, they do not damage them.[2]

You can judge which of our trees are hardy and which very hardy simply by knowing their geographic ranges. Only the very hardy ones can grow in the subarctic. A greater proportion of conifers than of hardwoods are very hardy. This explains why there is a greater variety of conifers in the northern forests. The conifers are Jack Pine, Tamarack, White and Black Spruce, and Balsam Fir. Among the hardwoods, there are only three very hardy trees; they are Paper Birch, Trembling or Quaking Aspen, and

Balsam Poplar; but there are, as well, a number of very hardy shrubs, especially among the alders and willows.

NOTES

1. M. B. Davis, "Quaternary History and the Stability of Forest Communities," *in Forest Succession: Concepts and Applications*, D. C. West, H. H. Shugart, and D. B. Botkin, eds. (New York: Springer-Verlag, 1981).

2. M. J. Burke et al., "Freezing and Injury in Plants," *in Annual Review of Plant Physiology*, vol. 27, pp. 507–28, 1976.

Chapter 2

Ten Groups of Conifers

The trees to be considered in this book are the evergreen cone-bearing trees of northern North America, plus the larches (which are not evergreen) and the yews (which don't bear cones) and the shrub junipers and yews (for not all junipers and yews are trees). What it is that unites these plants in the botanical sense will be explained in Chapters 4 and 5. That they form a cohesive group, resembling one another much more closely than any of them resembles the broad-leaved trees, nobody would deny. The resemblances among some of them are very close, making it necessary to examine them carefully if they are to be correctly identified.

Our first task is to consider the various groups of conifers, and how they can be told apart, but first a few paragraphs on the naming of plants will be helpful.

The English names of the various groups of conifers are familiar to most people: the pines, the spruces, the firs, and so on. Each group is technically known as a *genus* (plural *genera*), and belonging to each genus are one or more *species*, which are described as members of the genus. (The fact that the word species is the same in the plural as in the singular causes frequent misunderstandings: In writing, relief is obtained by using the abbreviations sp for the singular and spp for the plural of the single word species.) Thus every species belongs to a genus, and often two or more species belong to the same genus. For example, Lodgepole Pine and Red Pine are two different pines; in other words, they are two different species of the pine genus.

The scientific Latin name for every species of organism—and that includes, of course, every species of coniferous tree—is made up of two words: The first (always with a capital initial letter) is the name of the genus; the second (always with a lowercase initial letter) names the species within the genus. For example, all pines have the generic name *Pinus*; the specific names of two of the pines, Lodgepole Pine and Red Pine, are *Pinus contorta* and *Pinus resinosa* respectively. If several species in the same genus are mentioned, the name of the genus is written in full only the first time; after that its initial suffices. Thus we would write *Pinus contorta* and *P. resinosa*.

In formal scientific writing, the name (usually abbreviated) of the scientist who first described and named the species is added at the end of the two-part Latin name. Thus the full name of Lodgepole Pine is *Pinus contorta* Dougl. Here Dougl. is short for David Douglas, the Scottish explorer and plant-hunter who traveled and collected in British Columbia and the western United States in the 1820s and 1830s. Plant names are always printed with the Latin part in italics, and the author's name, when it is given, in roman type.

Sometimes two trees in different genera have the same specific name; for instance, the Western Larch is *Larix occidentalis* and the Eastern Arborvitae is *Thuja occidentalis*. Possession of the same specific name does not mean that the two trees are related any more than possession of the same first name means that two people are related. The specific name merely singles out and describes a particular member of a genus. In the examples above the descriptions are a bit ambiguous. The specific name *occidentalis* (western) applied to larch means the species is limited to western North America; the same descriptive name applied to arborvitae means that it grows in the western as opposed to the eastern hemisphere.

Now consider the ten genera of northern conifers. They are the pines (*Pinus*), the larches (*Larix*), the spruces (*Picea*), the hemlocks (*Tsuga*), the Douglas-firs (*Pseudotsuga*), the firs (*Abies*), the arborvitaes (*Thuja*), the false-cypresses (*Chamaecyparis*), the junipers (*Juniperus*), and the yews (*Taxus*).

For the moment, we consider only the characteristics of each genus that single it out from all other genera, in a word, the *diagnostic* characters. These are the bare minimum of characters that you must know in order to recognize a coniferous genus with absolute certainty. There is really very little to memorize.

THE PINES. In the pines, and only in the pines, the leaves are truly needle-shaped and, in the North, they grow in bundles (fascicles) of at least two and at most five needles, as in Figure 2.

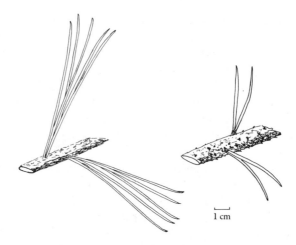

Figure 2. Pines.

THE LARCHES. The next instantly recognizable genus is *Larix*, the larches. They are the only deciduous conifers in the North. In summer the leaves, which grow in tufts from dwarf, stubby twigs, are soft to the touch and apple-green. In the fall they turn deep yellow or golden, and they are shed before winter. In its leafless winter state, a larch tree is still unmistak-

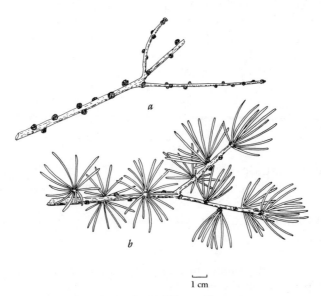

Figure 3. Larch. (*a*) Winter; (*b*) summer.

able and impossible to confuse with a broad-leaved tree. A larch tree as a whole has the typical pyramidal, Christmas tree shape of a conifer. In addition, the twigs and branches are covered with very distinctive knobbles. These are the dwarf "spur" twigs that bear the leaves in summer.

Now we come to a group of genera that are not so easy to tell apart. These are the conifers with short, narrow evergreen leaves that grow singly on the twigs (Figure 4). The leaves may spread out all around a twig, making it look like a bottle brush, or they may lie in a horizontal plane, projecting to left and right of the twig. Either way, they grow singly, not in fascicles as pine needles do.

These leaves are often called "needles," but the name is not quite so appropriate as it is for the leaves of the pines. The question now arises: How can we tell these genera apart? There are five genera to consider plus one juniper species (Common Juniper, *Juniperus communis*).

COMMON JUNIPER. We single out Common Juniper first, as it is easy to recognize. It is a shrub, not a tree. The prickly leaves are very sharp-pointed (awl-shaped), and are often curved so as to be almost clawlike. Sometimes the shiny lower surface is more bronze than green. There is a conspicuous white line down the center of the upper surface, which appears in the enlarged drawing (Figure 5*b*). As can also be seen in the drawing, the leaves are in whorls of three around the twig; however, the leaves are easily knocked off, and to find them in trios may take a lot of searching.

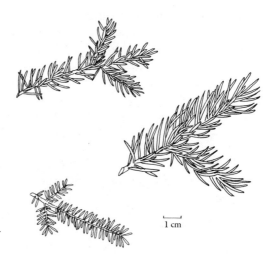

Figure 4. Different kinds of needle-leaves.

1 cm

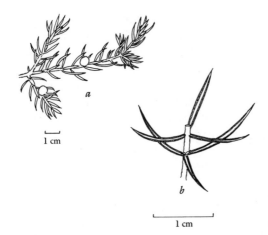

Figure 5. Common Juniper.

The female plants bear pea-sized "berries" with a blue bloom; two are shown in Figure 5*a*. These are not true berries but cones in which the scales have grown together and become fleshy. Because the sexes are separate in junipers, berries will be found only on female plants. They are useful as a diagnostic character nevertheless, as Common Junipers seldom grow singly; wherever the shrubs grow, there are likely to be several of them and some, at least, will be female and berry-covered.

The five genera now to be considered are the spruces (*Picea*), the firs (*Abies*), the hemlocks (*Tsuga*), the Douglas-firs (*Pseudotsuga*), and the yews (*Taxus*).

THE SPRUCES. The spruces are easy to recognize because their leaves (needles), instead of being more or less flat, are square in cross section like a wooden match; and like a match, a spruce needle will roll between your thumb and forefinger. This method of recognition works everywhere except on the West Coast, where Sitka Spruce (*Picea sitchensis*) grows; its needles, though four-cornered, are too flat to roll. Therefore on the West Coast (and only there) it is necessary to use a second characteristic of spruces to recognize them with certainty. If you look at the dead branches and twigs that are nearly always to be found low on a spruce tree, you will see that they are very rough, being covered with numerous small, hard "pegs"; these are persistent woody leaf stalks that remain on the twigs after the leaves have died and fallen. (Be sure that the twig you examine is not so old that the bark has peeled off.) Not only the pegs, but also the deep pattern of grooves on the twig are characteristic. Thus you can tell immediately whether a tree is a spruce by examining its dead twigs. The method works everywhere. Figure 6*a* shows a typical dead twig; note the charac-

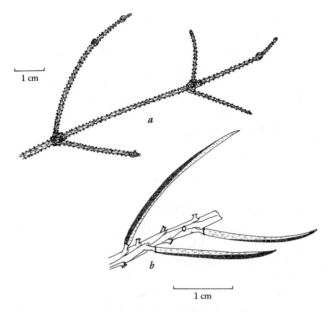

Figure 6. Spruce.

teristic thickenings, some of them where the twig branches. Figure 6*b* is an enlarged drawing showing part of a twig with three living leaves still attached to the projecting pegs.

THE FIRS. The firs (*Abies*) are also easy to recognize. They have two dependable diagnostic characters.

First, the bark: Except when it is very old, it is conspicuously smooth, and pocked with resin blisters (Figure 7*a*); when one is cut open, the

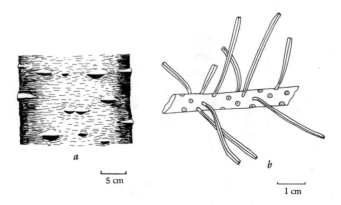

Figure 7. Fir.

excessively sticky resin pours out. Because of the way trees grow, continuously adding new growth at the top, young bark can be found at the top of any tree; only the bark at the bottom of the trunk is as old as the tree itself. Old fir bark does, eventually, become fissured (Figure 37c).

Second, the twigs: Look at the base of a leafy twig where the old leaves have fallen. The leaf scars are perfectly circular and flat (or even slightly recessed), making the leafless twig smooth to the touch (Figure 7b). This texture is in marked contrast to the rough leafless twigs of spruce and the slightly roughened twigs of hemlock.

THE HEMLOCKS. A hemlock is most easily recognized by standing back and looking at the tree as a whole. It is far more graceful, and more lacy and feathery, than the other conifers. The reason is that the leaves are small and the outer ends of the branches are slender, flexible, and drooping. The most unmistakable characteristic of a hemlock is its dangling leader. The *leader* (leading shoot) is the uppermost part of the stem of a tree, above all the side branches.

The leader of a young hemlock is pliable, and curves over to give the tree a gracefully nodding tip (Figure 8b). In an old tree, the graceful curve usually becomes an angular twist or a bend. Each leaf tapers at the base to a short, slender stalk that makes a sharp angle with the leaf's midrib. The leaves are of a variety of different lengths (Figure 8a)—anywhere from 0.5 to 2 cm (¼ to ¾ in.)—all mixed together (this characteristic is much less pronounced in mountain hemlock). The leafless twigs (Figure 8c), though not smooth, are not nearly so rough as spruce twigs.

THE DOUGLAS-FIRS. The next trees to consider are the Douglas-firs. They are found only in the West, from the Rockies to the Pacific. They

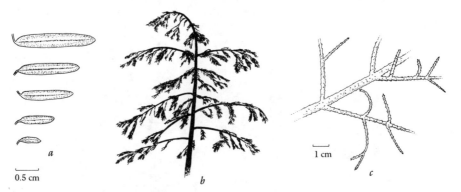

Figure 8. Hemlock. (*a*) Five leaves from the same twig; (*b*) the leader and uppermost branches; (*c*) leafless twig.

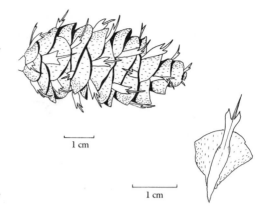

Figure 9. Douglas-fir cone and bract.

have unique cones. For most genera of conifers, cones are not the best diagnostic feature. At the time when we want to identify it, there may be none on a tree, or, if there are, they may be so high as to be out of reach and impossible to see clearly. But Douglas-firs have plenty of cones for many months of the year, and on trees growing in the open they are found on the lower as well as on the upper branches (unlike spruce and fir cones, which are usually confined to the tops of the trees).

After the cones have fallen, they are easy to find on the ground under the tree. Their distinctive feature is the three-pronged bract (Figure 9*b*) that protrudes from the cone behind every scale (the distinction between scales and bracts is explained in Chapter 4). As no other conifer has cones like these, they provide a sure means of identification.

Without the cones it would be possible to confuse Douglas-fir with fir (not, of course, in the East where, except in plantations, Douglas-fir does not grow). In very young Douglas-fir the bark is smooth and resin-blistered like that of a fir. Also, as in Figure 10, the leafless twigs are fairly smooth.

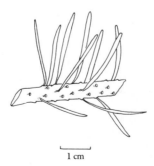

Figure 10. Douglas-fir.

But if any doubts arise, they can usually be laid to rest by examining the buds. Those of the firs are rounded (Figure 11*b*), and have a sticky coat of resin. Those of Douglas-fir are pointed (Figure 11*a*), and are dry and papery to the touch. The next season's buds are large enough to be seen at any time of year except just after the opening of the current season's buds and the unfolding of the new leaves within them. The twigs in the drawing were collected at the end of July. The buds are most easily found by turning the twigs over and looking at their lower sides; the drawings are of the lower sides of the twigs.

THE YEWS. The last of the needle-leaved genera to consider is the yews (*Taxus*). There are two species, one in the East and the other in the West. The eastern yew is a shrub that never reaches tree size. The western yew grows to the size of a small tree [rarely more than 8 meters (25 ft) tall] when it grows in moist, shady forest environments; but in less hospitable sites it, too, grows no larger than a shrub. In yews, as in junipers, the sexes are separate, and when a female plant is covered with "berries" it is unmistakable. The berries are soft, fleshy, coral-red "cups," open at the top; a single, hard, dark-green seed can be seen within each cup. The seeds are dangerously poisonous, but the red flesh is sweet and juicy and harmless to most people. These brightly colored fruits are not true berries in the botanical sense; they are highly modified cones.

Male yews, or females without berries, are not quite so easy to recognize, but the leaves are distinctive enough once you know what to look for. They are flat, dark green above, and noticeably paler below; and they taper abruptly at the tip to a very slender point, a kind of leaf tip that is technically known as *mucronate*.

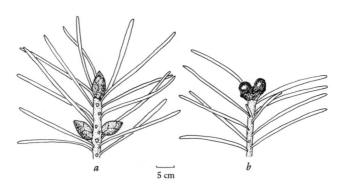

Figure 11. (*a*) Douglas-fir; (*b*) Fir.

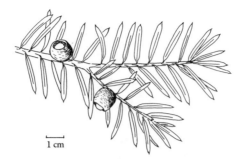

Figure 12. Yew.   1 cm

THE JUNIPERS, ARBORVITAES, AND FALSE-CYPRESSES. Now we come to the three genera of conifers with overlapping scalelike leaves, as closely pressed to the twigs as shingles to a roof. These are the arborvitaes (*Thuja*), the false-cypresses (*Chamaecyparis*), and all the junipers (*Juniperus*), except Common Juniper, which was described before. *Thuja* is sometimes spelled *Thuya* and so pronounced. Note, also, that *Thuja* and *Chamaecyparis* have better-known English names than those given here. There is more to be said about names at the end of this chapter.

The junipers are easily distinguished from the other two genera by the fact that their branches are bushy instead of forming flat frondlike sprays. Figure 13 shows a branch of juniper on the left and of arborvitae on the right. When in doubt, ask yourself if you could spread out an armload of twigs to form a flat carpet on the ground. If you could, the tree is an arborvitae or a false-cypress; if not, it is a juniper.

Distinguishing between an arborvitae and a false-cypress in regions

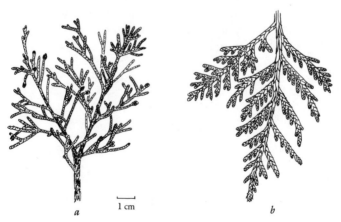

Figure 13. (*a*) Rocky Mountain Juniper; (*b*) Giant Arborvitae.

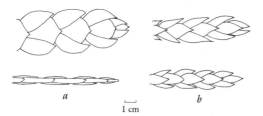

Figure 14. (a) Arborvitae; (b) False-cypress.

where their geographic ranges overlap requires a closer look. But first, consider the ranges. Except out West, within 200 km (150 mi) of the Pacific coast, any scale-leaved conifer with flat fronds found growing wild is an arborvitae. In the area where false-cypress is a possibility, it can be distinguished from arborvitae by a close look at the twigs. In both genera the scalelike leaves face each other in pairs, each pair being at right angles to the pair above and below it. As a result, the leaves form four longitudinal rows up and down the twig. In false-cypress the four rows are all alike (or nearly so), and although an individual twig is somewhat flattened, it has much the same pattern from whichever side it is viewed.

In arborvitae, on the other hand, the leaves are not all alike and the twig is clearly flat. The two flat surfaces of a twig are formed by facing pairs of flat leaves, and the two edges of a twig by facing pairs of folded leaves. Figure 14, showing the front and side views of both kinds of twigs, should make the matter clear.

Arborvitae and false-cypress can also be told apart by their barks. Although a great many trees have barks that are not especially distinctive, and are exceedingly hard to describe, the barks of these two genera are strikingly different. That of arborvitae is stringy, and it can be pulled off in long, thin, pliable fibers; it is brown. The bark of false-cypress is light gray, and it forms brittle rectangular flakes that peel off revealing (especially in young trees) a fresh, dark-red layer.

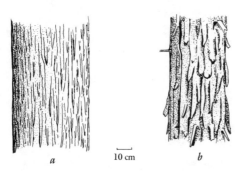

Figure 15. (a) Arborvitae; (b) False-cypress.

To summarize what has been said above on recognizing the ten genera of conifers in our area, it will help to look at a diagnostic "key." The key is shown below. Unless we understand how a key works, it seems pure nonsense—a list of pairs of contradictory statements.

Key to the Genera

1. Leaves not evergreen ....Larch
1. Leaves evergreen...2
  2. Leaves needlelike..3
  2. Leaves scalelike 10
    3. Leaves in fascicles (bundles) of two or more..................Pine
    3. Leaves growing singly 4
      4. Trees 5
      4. Shrubs 9
        5. Leaves square in section (except Sitka Spruce), growing
          from woody pegs...Spruce
        5. Leaves not square in section 6
        6. Leaf scars flat, circular; buds round, resin-coated ......Fir
        6. Not as above..7
          7. Cones with trident-shaped bracts;
            buds pointed ....Douglas-fir
          7. Not as above..8
            8. Trees with drooping leader; leaves not more than 2
              cm long, with thin stalk ....Hemlock
            8. No drooping leader; leaves dark above, pale below,
              up to 2.5 cm long.....................................Yew
        9. Leaves stiff, curved, sharp, with
          white line.......................................Common Juniper
        9. Leaves flat, dark green above, pale green below ....Yew
  10. Branchlets bushy...Juniper
  10. Branchlets forming flat sprays.....................................11
    11. Pairs of scale leaves alternately flat and folded.....Arborvitae
    11. All pairs of leaves partly folded ....False-cypress

The contradictions, however, are the secret of how it does work. Suppose you have a conifer to identify and are using the key to guide you. Read the first two sentences of the key, each labeled with a 1 on the left. The two sentences are

        1. Leaves not evergreen..............Larch
        1. Leaves evergreen.........................2

Obviously, as the sentences contradict each other, only one of them can be true. Not only that, one of them *must* be true. You now decide which of the two *is* true and look to the right of the true statement. Either you see a name, which is the answer you are seeking; or you see a number, in this case a 2. Suppose the tree is evergreen and you are therefore led to the 2. This means you should next read the pair of sentences labeled on the left with 2. They are

> 2. Leaves needlelike...............3
> 2. Leaves scalelike 10

If the leaves of the specimen are needlelike, you now read the two sentences labeled 3; if the leaves are scalelike, you read the two sentences labeled 10. And so on.

A key is simple to understand and, in principle, easy to use. But it is possible to be led astray, and it is worth considering the difficulties that may crop up when we construct or use a key.

First, the person who constructs a key must be sure that a statement intended to be true of any particular group of plants is true of all the plants in the group. This means that it is sometimes necessary to avoid mention of a conspicuous, easily observable character found in only some members of a group in favor of a less easily observable character found in every member. For example, consider the hemlocks. In nearly all the hemlocks, the leaves of the same age on a single twig vary considerably in length; short ones and long ones are mixed together. This variability would provide a useful diagnostic character for the hemlocks were it not for the fact that it is scarcely noticeable in mountain hemlock. Therefore the character must be ruled out as a dependable one for singling out the hemlocks; it will not work well for all of them.

Likewise, the blue berries on female junipers are not mentioned in the key because, obviously, they are of no help in recognizing the males. For identifying the spruces, the square shape of the leaves in cross section (except in Sitka Spruce) is so very useful as a diagnostic character that it is worth breaking the rule for; of course an additional character (the peglike leaf bases) has to be mentioned to take care of the exceptional Sitka Spruce with almost flat leaves.

The next requirement of a key is that it should be usable at any time. A unique and conspicuous character of the firs is that their ripe cones stand upright on the branches. But they are only visible for part of the year and for that reason their uprightness is not mentioned in the key.

Ideally, all the pairs of contrasted statements in a key should leave the user in no doubt about which of the two to choose. Sometimes this ideal state of affairs is hard to achieve. For instance, look at the pair of statements labeled 8. They mention leaf length (less than 2 cm in hemlock and up to 2.5 cm in yew) as a help, but leaf length alone could be misleading; thus a tree with leaves 2 cm long could be either a hemlock, or a yew with rather small leaves; therefore the key must mention other differences as well.

The last point to mention about keys is that the user should always know the limits of the geographic region that the key is intended to cover. No key can be counted upon to work outside its own region. For example, in this key, which applies to conifers in North America north of 45° north latitude, the pines are recognized as all those trees that are evergreen and have leaves (needles) in fascicles of two or more. If we used the key in Nevada, say, and were trying to identify a Single-leaf Pinyon Pine, the key would be misleading; at the pair of statements labeled 3 we would be led away from the pines and would reach an erroneous identification. This would not be the fault of the key but of the user, for trying to use it outside its proper region.

Not every coniferous tree you come across is easy to identify, of course, even with a key. A tall tree in dense forest, such as the West Coast rain forest, can present problems. Its lowest branches may be far overhead, making twigs, leaves, and cones unreachable. There may be no young bark in sight, and the old bark will often be overgrown with moss. And because it is surrounded by other trees, it will not have developed the characteristic shape of open-grown specimens of its species. These difficulties disappear as you become familiar with the trees of a region; and familiarity is soon acquired if you first identify a number of "easy" specimens and gain a feel for each species.

This is the place to say a little more about the names of trees. Botanists are sometimes accused of being unnecessarily pedantic for calling trees by their scientific Latin names instead of their common English names. The reason for preferring the scientific names is that they are unambiguous. A name such as *Thuja plicata*, say, means one thing and one thing only; to call it red cedar, as many westerners do, invites confusion with an easterner's idea of a red cedar, which is the eastern tree-sized juniper, *Juniperus virginiana*, not even a member of the same genus. Moreover *Thuja plicata* means the same thing all over the world to all botanists, no matter what their mother tongue.

The name cedar is a source of endless confusion. Ideally it would be used only for the true cedars (genus *Cedrus*), none of which grow wild in North America. Where the climate is mild enough, species of *Cedrus* are often planted as ornamentals, but they are found wild only in the Himalayas, Asia Minor, and Mediterranean Africa. In North America the name cedar is used, confusingly, for several unrelated trees: *Thuja plicata* (western red cedar), *Juniperus virginiana* (eastern red cedar), *Thuja occidentalis* (eastern white cedar) and *Chamaecyparis nootkatensis* (Alaska cedar or yellow cedar), to name only those growing in our area (there are several more, south of the area). Some authors, it is true, try to forestall misunderstanding by linking one of the distinguishing adjectives more closely to the word cedar; instead of red cedar, one can write red-cedar or even redcedar. But it seems simplest and least muddling to use the English names for the different genera that have already been introduced: arborvitae for *Thuja*, juniper for *Juniperus*, and false-cypress for *Chamaecyparis* (not cypress, although it is often called that, because cypress should be restricted to the genus *Cupressus* which does not grow in our area).

Using a single English name, such as cedar, for trees belonging to several different genera is only one form of the naming muddle. The muddle can be (and is) reversed. That is, a single species of tree may be called by many English names. A good example is the species *Pseudotsuga menziesii*. Most people call it Douglas-fir and if everybody did so no problem would arise. But in some districts it is called Douglas spruce or bigcone spruce, and in the form of lumber it is often sold as Oregon pine. Even the standard name, Douglas-fir, is not altogether unobjectionable. Often the hyphen is omitted. It shouldn't be, for if it is, it implies that Douglas-fir is simply another of the firs, a member (along with Balsam Fir, Alpine Fir, Grand Fir, and so on) of the genus *Abies*. Even worse, Douglas-fir is frequently simply called fir for short, with the odd result that the habit has grown up of speaking and writing of *Abies* species as true firs. This is done even in the technical forestry literature, to show when the trees meant are truly firs (*Abies*) as distinct from the false variety (*Pseudotsuga*), whose Latin name when translated actually means "false hemlock."

Other naturalists, most notably birders and insect specialists, have managed to standardize the English names they use as informal alternatives to scientific Latin names. It will be a great step forward when plant lovers do likewise. English and Latin names are used together throughout this book; wherever a selection among several English names had to be made, the choice was made so that there should be a one-to-one correspondence between the English and the Latin versions: each English name refers to one, and only one, botanical genus.

Chapter 3

The Species

The previous chapter dealt with the different genera of conifers and how, with the barest minimum of memorized knowledge, to tell them apart. This chapter describes the 10 genera in our area in more detail, and explains how to recognize the different species within each genus.

There is, of course, plenty to be said about the cones of conifers. All conifers bear two kinds of cones: seed-bearing, or female, cones and pollen-bearing, or male, cones (the exception is the yews, in which the seed is borne in a fleshy "berry" rather than in a cone). In pine, larch, spruce, hemlock, fir, Douglas-fir, arborvitae, and false-cypress, cones of both sexes are found on the same tree. In the junipers and yews, the sexes are separate; in these two genera a single tree (or shrub, as the case may be) is either male or female but not both.

In all genera, the pollen-bearing cones are small and short-lived. They are most noticeable in spring in the short period during which they grow, expand, open, and shed their pollen. Soon afterward they wither and most of them drop off. Further consideration of the pollen-bearing cones is deferred to Chapter 4, where the reproduction of conifers is described. In this chapter all references to cones are to the seed-bearing kind, which are the "ordinary" cones to be seen on the trees for many months of the year and found scattered on the forest floor. They will be referred to as cones throughout this chapter, without the adjective seed-bearing which is to be understood.

THE PINES

The genus *Pinus* is the largest conifer genus in our area, in the sense that it is represented by more species (nine) than any other genus. The pines differ from most other genera in that the cones take 2 years to reach maturity (the exceptions are false-cypress and many of the junipers). At most seasons we can therefore find unopened cones of two sizes on a pine tree: small, first-year, immature cones and large, second-year, ripe or ripening cones. In early spring it is possible to find three sizes: newly formed cones awaiting pollination, 1-year-old cones, and just-matured 2-year-old cones. Ripe cones may either open, shed their seeds, and then drop off; or they may remain on the tree for many years, closed up with the seeds inside. Figure 16 shows first- and second-year cones of Lodgepole Pine (*Pinus contorta*), both drawn to the same scale. The leaves have been removed so that the cones are not concealed.

We now consider how to tell the nine species of pines in our area apart. The task is not difficult. The first point to notice is that there are two groups of pines, the so-called hard pines and soft pines. The names refer to the hardness of the wood, a quality of little interest to a field naturalist. However, the separation into two groups is not an arbitrary one devised for the convenience of carpenters and other wood users. The hard and soft pines are distinctly different in the botanical sense. These are the differences:

1. The hard pines have needles in fascicles of two or three, whereas in soft pines the needles are in fives.
2. The needle of a hard pine has two "veins" (fibrovascular bundles) and the needle of a soft pine only one. The difference is easily seen by slicing through a needle with a razor blade and examining the cross section with a strong (× 10) hand lens. The shape of the leaf in cross section is also

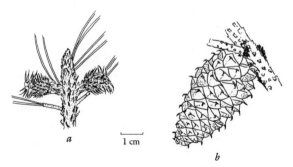

Figure 16. Lodgepole Pine cones. (*a*) First year; (*b*) second year (mature).

distinctive, as shown in Figure 17. The leaves of soft pines are triangular in section. Two different shapes occur in the hard pines, but there is always a curved, semicircular side.

3. In all pines, each young fascicle of needles has a papery sheath of bud scales wrapped around the base. In the hard pines the sheath remains permanently in place; in the soft pines the sheath is shed soon after the needles are full grown and is absent from mature fascicles (see Figure 17).

4. In the hard pines the scales of the cones usually have a sharp prickle on the back (as in Figure 16*b*). The cone scales of soft pines usually lack prickles. But note the word "usually." Prickliness or the lack of it is too inconsistent to serve as a diagnostic character. For instance, the cones of Red Pine (one of the hard pines) lack prickles, and those of Jack Pine (another hard pine) sometimes have minute prickles but are usually smooth (see Figure 20*a*).

In any case, to tell hard pines and soft pines apart, we have only to count the needles in the fascicles and remember that the soft pines have five and the hard pines fewer than five.

Of the 9 species of pine in our area, 5 are hard pines and 4 are soft pines. Of the hard pines, 3 are two-needle pines and 2 are three-needle pines. Of the two-needle pines, 2 are eastern and 1 a western species. All this information is gathered together systematically in the key at the end of this section. A key is a useful device for summarizing dry facts compactly, as well as being the surest guide to correct identification.

For the moment, we consider the three species of two-needle pines.

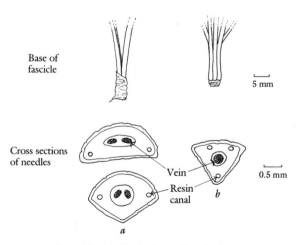

Figure 17. (*a*) Hard pines; (*b*) soft pines.

They are Red Pine and Jack Pine in the East, and Lodgepole Pine in the West. First, the Latin names: Red Pine is *Pinus resinosa*. Jack Pine is *Pinus banksiana* and is named after Sir Joseph Banks (1743–1820), one of the greatest of the many great explorer-naturalists of the 18th and 19th centuries, who carried out a botanical exploration of Newfoundland and later, aboard *Endeavour*, accompanied Captain James Cook on the first of his round-the-world voyages. Lodgepole Pine is *Pinus contorta*, but the name is not a good description of the shape of the tree as a whole, except for those growing within a short distance of the Pacific coast, which belong to a special, untypical group and are indeed contorted (see Figure 22*c*). Most Lodgepole Pines have tall, slender trunks as straight as ramrods.

As the eastern two-needle pines are entirely unlike each other, there is no risk of confusing them. The needles of Red Pine are 10 to 15 cm (4 to 6 in.) long while those of Jack Pine are 6 cm (2½ in.) at most and usually less. There is nothing particularly noteworthy about the cones of Red Pine except that when they drop from the tree they often leave a few basal scales behind, as in Figure 18.

Overall, the Red Pine is a colorful tree; its trunk is a deep, rich salmon color and its broad, oval crown a bushy mass of shining, dark green needles. The species does not occur west of Lake Winnipeg and its tributary the Red River (the boundary between Minnesota and North Dakota).

Now for Jack Pine and Lodgepole Pine: They are closely related to each other and are able to hybridize. However, there is no risk of confusing them in most of our area as their geographic ranges scarcely overlap. Any two-needle pine to the west of the dash-dot line in Figure 19 is sure to be a Lodgepole Pine and any to the east of it, if its needles are less than 6 cm (2½ in.) long, is a Jack Pine. Two-needle pines in the cross-hatched

1 cm

Figure 18. Red Pine. Rosette of basal scales left by a fallen cone.

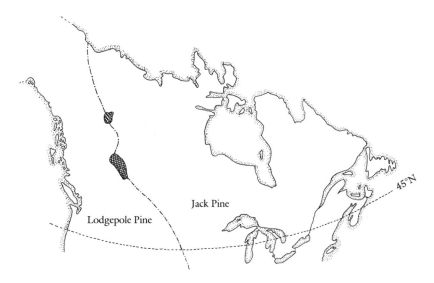

Figure 19. Boundary between Lodgepole Pine and Jack Pine ranges.

regions where the ranges of the two species overlap are quite likely to be hybrids and thus cannot be said to belong to either species. The hybrids have characteristics intermediate between those of the parent species, and the latter are not very different even when genetically pure.

The map differs from ordinary botanical geographic range maps in not showing, as shaded patches, regions where the species concerned are certain to be found. Instead, what it shows is a dividing line which separates the two species, together with shaded patches where clear division is impossible. The map does not imply that Jack Pine, for example, is found everywhere east of the line; what it does imply is that *if* a short-leaved, two-needle pine is found east of the line, *then* it is a Jack Pine.

Because they are closely related, Jack Pine and Lodgepole Pine have very similar cones and needles. In both species the needles are short [about 4 cm (1½ in.) on average], sharp-pointed, stiff, somewhat twisted (more so in Lodgepole) and yellowish green. The cones of the two species are also rather similar as shown in Figure 20. Both species have a tendency for the cones to be unsymmetrical and curved, and the tendency is more pronounced in Jack Pine than in Lodgepole. The prickliness of the cones is much more pronounced in Lodgepole. In Jack Pine, even though it is one of the hard pines, the cones are often smooth, and prickles, when there are any, are tiny. The greatest contrast between the two species is in the way the cones are attached to the branches. In Jack Pine the cones point

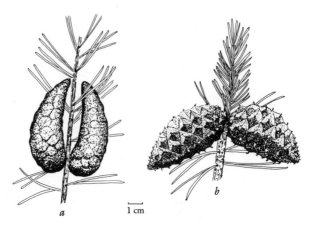

Figure 20. (*a*) Jack Pine; (*b*) Lodgepole Pine.

forward, toward the tip of the branch, whereas in Lodgepole they point backward.

The two species are similar in another respect. A large proportion of the mature cones remain on the trees, tightly closed, for as many as 20 or more years, with viable seeds inside. Such late-opening cones are known as *serotinous*, from the Latin *sero*, late. As a result, Lodgepole and Jack Pine trees are usually laden with cones—the accumulated cone crops of a long succession of years. Some cones persist for so long that the growing branch that bears them expands to engulf them; thus we may find healthy cones almost embedded in a branch. From time to time a few old cones open and shed their winged seeds, so that a mixture of open and closed cones will be found on a tree at any season, but the majority of the cones remain closed until a fire consumes the forest. Then the cones open because of the heat.

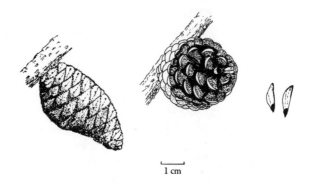

Figure 21. Jack Pine cones and seeds.

Even though a forest fire, especially a fierce one, must inevitably destroy a great many cones and seeds, there are always some survivors—often huge numbers of them—and these ensure the growth of a new forest to replace the burned one. There is more to be said on this topic in Chapter 8.

Although they are closely related, and in spite of their many resemblances, Lodgepole and Jack Pines look quite different from a distance. Not only that, Lodgepole Pines occur in two "forms," with strikingly different shapes. Jack Pine is a fairly short tree, 15 m (50 ft) tall on average, and it often appears rather threadbare; dead or dying branches may remain protruding from the trunk, making the crown look open and sparse. Lodgepole Pine in the interior part of its range is a tall, slender tree, especially when, as is usually the case, it grows in dense, pure, even-aged stands. When full grown, it averages 25 m (80 ft) in height. The coastal form of Lodgepole Pine is altogether different and is frequently called by a name of its own, shore pine. It tends to be short [about 10 m (30 ft)], is often twisted or contorted, and has one or more dense, smooth-topped crowns shaped like the caps of mushrooms. Figure 22 shows typical shapes, but it should be remembered that tree shapes are variable and that atypical specimens are plentiful.

Next, the three-needle Hard Pines. There are only two of these trees in our area, Ponderosa Pine in the West (also known as yellow pine), and Pitch Pine in the East.

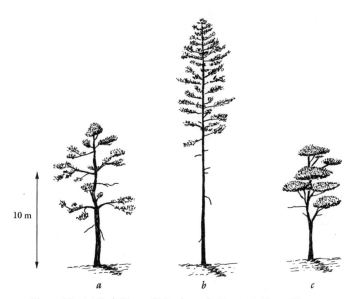

Figure 22. (*a*) Jack Pine; (*b*) Lodgepole Pine; (*c*) Shore Pine.

Ponderosa Pine (*Pinus ponderosa*) is a tree well known to all westerners. It forms open, uncrowded, parklike forests in the dry interior valleys of British Columbia, Washington, Idaho, and Montana, below the zone of dense forest on the mountainsides; it also grows, with Douglas-fir, within the mountain forests. In Canada, it is found west of the continental divide only (that is, in British Columbia); it is absent from the eastern foothills, in Alberta. In the United States it is found as far east as the Dakotas (and south of our area in Nebraska); its eastern limit is the Missouri River. Ponderosa Pines are big trees, up to 50 m (170 ft) tall, with long, dark needles and big, prickly cones. When the cones fall, soon after reaching maturity, they often leave a few basal scales behind as do the cones of Red Pine. Typical Ponderosa Pines, growing in dry soils, have very characteristic bark.

It is cinnamon- to rust-colored, and flakes off in pieces that resemble the pieces of a jigsaw puzzle. A layer of these flakes will usually be found around the base of a tree. Ponderosa Pines growing in damper soils do not develop this distinctive type of bark; their bark is dark, furrowed, and not especially conspicuous. Such trees are sometimes treated as a distinct form of Ponderosa Pine and are referred to as blackjack pines.

The last of the hard pines to consider here is Pitch Pine (*Pinus rigida*); it occurs only in an extremely small part of our area, the St. Lawrence Valley upstream of Montreal, but it is a common tree farther south, in New England and in the Appalachians as far south as the Carolinas. Like Jack Pine, it manages to grow in dry, sterile sites, where it tends to be small and scrubby and to have rather sparse foliage. Again like Jack Pine, it has serotinous cones that remain on the branches for years; the cones have sharp prickles. Two interesting features of Pitch Pines are the way in which tufts of needles grow on the trunk, and the fact that new stems sprout from

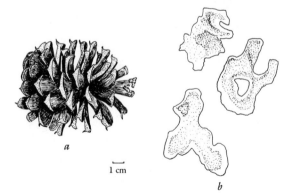

a

1 cm

b

Figure 23. Ponderosa Pine.
(*a*) Cone; (*b*) detached bark flakes.

the stump after a tree has been felled or burned. The only other conifers in which this happens are the yews.

Now to consider the soft pines, of which there are four species in our area, three in the West and one in the East. First it should be remarked that in the same way that pines as a whole can be divided into two subgroups, the hard pines and the soft pines (technically these are, respectively, the subgenus *Diploxylon* and the subgenus *Haploxylon*), so also the soft pines can be "subsubdivided" into several "subsubgroups." The four soft pines we need to consider belong to two of these subsubgroups, known as the white pines and the stone pines. They do not resemble each other at all closely, although all are five-needle pines. The white pines are tall, forest trees with soft, slender needles. The stone pines are often short and sprawling and more like shrubs than trees; they have to be, in order to survive in their dry, exposed, inhospitable habitats. Their needles are stiff, curved, and rather thick.

First, the white pines: Our area has two species called (rather uninterestingly) Eastern White Pine and Western White Pine. Their Latin names are, respectively, *Pinus strobus* and *Pinus monticola*. As their geographic ranges do not overlap, we can be sure about which of the two species a white pine belongs to.

The eastern species does not grow west of Lake Winnipeg and the Red River; the western species does not grow east of the continental divide.

Mature white pines have distinctive layered silhouettes; Figure 25 shows the well-known shape of an Eastern White Pine. One reason why it is so familiar to naturalists is that Eastern White Pines, when full grown, are the tallest of all the coniferous trees in their region, up to 50 m (170 ft) and more. Western White Pines, in contrast, do not stand out among their

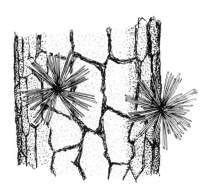

Figure 24. Pitch Pine (with tufts of needles on the trunk).

5 cm

2 m

Figure 25. Eastern White Pine.

equally tall forest neighbors. Even so, their crowns have the typical layered form, but it is usually narrower and more symmetrical. Because Western White Pines seldom tower over surrounding trees in the way that Eastern White Pines do, their upper branches cannot grow so long, and they are not exposed to strong prevailing winds that might make them unsymmetrical. They do stand out from the trees that surround them, however, in having cones so large that they are visible from a distance, especially when the top of a tree is seen in silhouette.

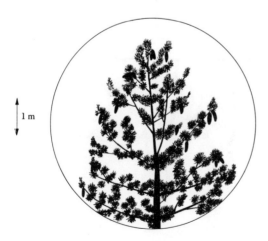

1 m

Figure 26. Western White Pine.

1 cm

Figure 27. Eastern White Pine.

The cones of both white pine species are narrowly cylindrical, up to 20 cm (8 in.) long in the eastern species and 25 cm (10 in.) in the western. They open and shed their winged seeds in the fall soon after reaching maturity, and the open, empty cones fall from the tree within a few months. The drawing shows a closed 2-year-old cone approaching maturity and an open cone from which the seeds have been shed.

The two stone pines in our area (they are the only stone pines in North America) are Whitebark Pine and Limber Pine. These names are merely translations of the Latin names: *Pinus albicaulis* and *Pinus flexilis*. Two of their shared characteristics, namely, their bushy, stunted shape (in their usual habitat) and their stout, stiff needles, have already been mentioned. Another peculiarity of the stone pines is that, unlike all other pines, their seeds do not have wings.

The two species, Whitebark and Limber, are very alike in all respects except for their markedly dissimilar cones. Specimens without cones are almost impossible to tell apart. The qualities described by their names are shared by both species: In both, the bark is smooth and almost white on young stems, although it becomes dark and fissured with age. Also in both, the branches are remarkably flexible; this enables them to bend without breaking and so survive the heavy snows and strong winds to which they are exposed in their mountain habitats.

Both species are restricted to the western mountains. Whitebark Pine has the larger range; it is found most often in exposed, rocky sites near timber line in the Rockies and the Cascades south of 55° north latitude. Dwarfed individuals are found above timberline, scattered in the alpine tundra. Limber Pine does not grow as far north and it does not occur west of the Rockies in our area. Besides growing in the same habitats as

1 m

Figure 28. Limber Pine.

Whitebark Pine, at the upper limit of trees, it also occurs in the eastern foothills of the Rockies where dry, stony ridges extend out into the plains.

The cones of the two species are strikingly different, not only in shape and size (as shown in Figure 29) but also in the way they shed their seeds. In Limber Pine, as in all other pines, the scales spread apart (that is, the cones "open") to release the seeds. In most pines this happens as soon as the seeds are ripe, but in the serotinous pines (Jack, Lodgepole, and Pitch pines) many years may pass before a mature cone, with ripe seeds inside, opens. In Whitebark Pine, however, the cone scales remain permanently closed. When the cones are mature, in early fall of their second year, they decay and disintegrate so that the seeds fall free (unless, as often happens, squirrels or birds have already broken them open).

Notice that in some Limber Pine cones the tips of the scales are bent back (or reflexed) as shown here, whereas in others they are not. But there is no risk of confusing Limber and Whitebark pines if cones can be found; the contrast between the cones in size and overall shape is consistent and ensures certain identification.

To conclude this section on the pines, here is a key to the nine species in our area.

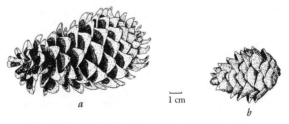

a   1 cm   *b*

Figure 29. (*a*) Limber Pine; (*b*) Whitebark Pine.

Key to the Pines

1. Two or three needles per bundle (hard pines)...........................2
1. Five needles per bundle (soft pines) 6
2. Two needles per bundle 3
2. Three needles per bundle..5
3. Western tree..Lodgepole
3. Eastern tree...4
4. Needles 10 cm or more in length ....Red
4. Needles 6 cm or less in length ....Jack
5. Western tree...Ponderosa
5. Eastern tree (St. Lawrence valley only).........................Pitch
6. Needles thick and stiff. Exposed mountain habitats.................7
6. Needles slender and soft. Tall forest trees............................8
7. Cones long (more than 7 cm), cylindrical...................Limber
7. Cones short (less than 7 cm), egg-shaped...............Whitebark
8. Eastern tree ....Eastern White
8. Western tree ....Western White

THE LARCHES

There are only three species of larch in our area, Eastern Larch or Tamarack (*Larix laricina*), Western Larch (*Larix occidentalis*), and Alpine Larch (*Larix lyalli*), which is also a western species.

The geographic range of the eastern species does not overlap that of the two western species. The dividing line is shown in Figure 30. It does not

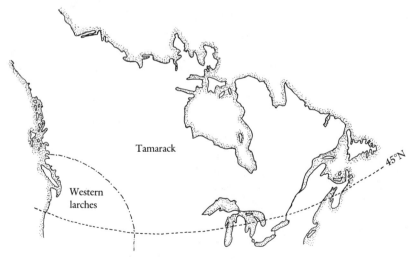

Figure 30. Boundary between the ranges of Tamarack and the western larches.

coincide with the continental divide; Tamarack is to be found west of the continental divide in the region of the big bend of the Fraser River, near Prince George, and Western Larch and Alpine Larch, although confined to the mountains, occur on both sides of the height of land but only south of about 52° north latitude.

There is no risk of confusing Tamarack with either of the western larches, even where their ranges come close, provided cones can be found. The cones of both western species have long, tapering, pointed bracts protruding behind the cone scales. In Tamarack, although bracts are present, they are shorter than the cone scales and are not visible. This contrast is shown in Figure 31. Moreover, Tamarack cones are only about half the size of Western Larch cones, and the cones of Alpine Larch are bigger still, up to 5 cm (2 in.) long compared with 1 cm (less than ½ in.) for Tamarack.

The three species also differ greatly in the size of the full-grown tree. The largest is Western Larch, a tall, stately tree that can grow to a height of 60 m (200 ft); second is Tamarack, which seldom exceeds 20 m (70 ft); the smallest is Alpine Larch, a small, often stunted, timberline tree, rarely over 10 or 12 m (30 or 40 ft) tall. The two western larches are usually quite distinct, although they grow in the same geographic region. They differ in size, as already remarked, but this distinction is not useful when we want to identify a young, partly grown specimen. They differ in habitat: Western Larch is typically found in valleys in the western mountains or on lower mountain slopes below about 1500 m (5000 ft) elevation; Alpine Larch is a subalpine species (indeed, it is often called subalpine larch) and is seldom found below 2000 m (7000 ft) elevation. Their young twigs are also different. Those of Alpine Larch, but not of Western Larch, are covered with a dense layer of white "wool" (tangled, matted hairs); this is most easily seen with a hand lens.

Although most larches in the region where Alpine and Western larches

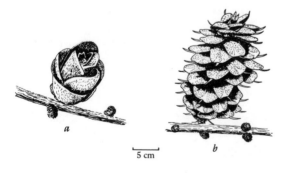

5 cm

Figure 31. (*a*) Tamarack; (*b*) Western Larch.

both occur are easily assigned to one or other of the two species, a few trees
with intermediate characters can be found and are probably hybrids.

THE SPRUCES

After the pines, which have nine species, the spruces in our area form the
next largest conifer genus, with six native species: White, Black, Red, and
Blue Spruce; Engelmann Spruce; and Sitka Spruce. Their Latin names are
Picea glauca, *P. mariana*, *P. rubens*, *P. pungens*, *P. engelmannii*, and *P.
sitchensis*.

It will help to sort them out geographically first. This is done in Figure
32, which is divided into seven regions labeled with numbers. The spruce
species to be found in each region are as follows:

1. White, Black, and Red spruces, and hybrids between Blacks and Reds.
2. White and Black spruces only.
3. White and Engelmann spruces and hybrids between them.
4. Engelmann Spruce only.
5. Hybrids of White and Engelmann Spruce and also Blue Spruce.
6. Sitka Spruce only.
7. Grassland; no spruces.

Now for the distinguishing characters of the various species, which we
consider region by region, starting in the east.

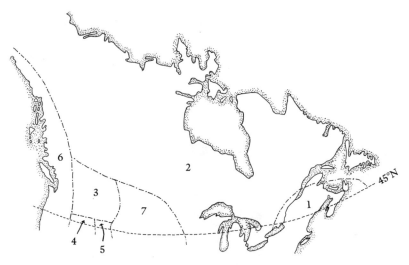

Figure 32. Regions of spruces.

As shown in the list, there are three species in region 1, the White, Black, and Red spruces. It is easy to separate White Spruce from the other two by examining young, fresh twigs of the current year. Those of White Spruce are hairless, smooth, and almost white; those of Red and Black spruces are covered with a dense fur of short, dark hairs (best seen with the aid of a lens) and appear rusty brown.

Distinguishing Red from Black Spruce is also straightforward provided cones can be found. Red Spruce has long, narrow cones whereas those of Black Spruce are short and almost spherical.

Spruce trees without cones are common, however. These are likely to be Red Spruce, as in this species the cones fall soon after they are mature. Black Spruce cones remain attached to the branches for years, releasing their seeds a few at a time. If cones cannot be examined, either because they are absent or out of reach, the two species are probably impossible to distinguish with certainty. They are capable of interbreeding, and where they grow together, hybrids with intermediate characters are likely to occur.

In region 2, which covers all of northern Canada except the West Coast, as well as the forests around the Great Lakes, the only spruces to be found are the White and the Black. A rough-and-ready method of distinguishing them at a distance is to note their shapes. Black Spruce has a narrow crown that is often densely bushy at the top; it differs in this respect from all the other spruces, which are very much alike in general outline and lack the distinctive terminal "blob." We cannot, however, rely on the blob as a dependable diagnostic character; not all Black Spruces have it.

In region 3 Engelmann Spruce appears. It is closely related to White Spruce, and as the two species are interfertile, hybrids with intermediate characters are common. Probably, over much of the region, hybrids out-number genetically pure trees of either species.

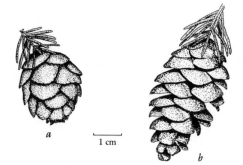

a   1 cm

b

Figure 33. (*a*) Black Spruce; (*b*) Red Spruce.

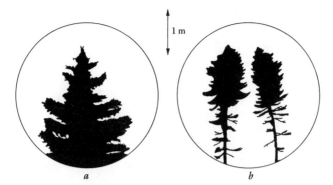

Figure 34. (*a*) White Spruce; (*b*) Black Spruce.

The cone of a genetically pure White Spruce differs from that of a genetically pure Engelmann Spruce in the way shown in Figure 35.[1]

The cone scale of a White Spruce has a smooth edge, and it is not much longer than the wings of the two seeds that it encloses. The cone scale of an Engelmann Spruce has a jagged edge (the technical word for "jagged" is *erose*) and is considerably longer than the seed wings. Even when a cone has shed its seeds, the shapes and sizes of the seed wings remain visible for a long time as discolored patches on the inner surfaces of the cone scales. So to decide whether a spruce in the western mountains is closer to a White or an Engelmann, we have only to find and examine a fallen cone. But except at the borders of region 3 it is unlikely to be a genetically pure member of either species.

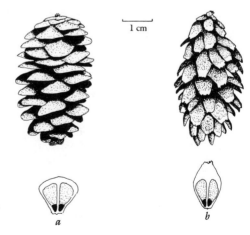

Figure 35. (*a*) White Spruce; (*b*) Engelmann Spruce.

Region 4, which is very small, is outside the geographic range of White Spruce and the only spruces found in it are Engelmanns.

Region 5 is like region 3 in containing hybrids of Engelmann with White Spruce, and in addition Blue Spruce is found there. Blue Spruce (also known as Colorado Blue Spruce) is, in its wild state, a tree of the high mountains. It occurs, though never abundantly, in scattered "islands" of high elevation in the Rocky Mountains, mostly south of 45°N. Within our area (north of 45°N) it can be found only in southern Montana, near the northeastern corner of Wyoming. In its usual frigid, windswept habitat, it is often rather stunted and seldom shows the silvery blue color that makes horticultural specimens so admired. In fact it can easily be confused with Engelmann Spruce with which it sometimes grows. The best way to distinguish between the two species is to chew a few leaves; those of Blue Spruce are sharply acid (hence the name *pungens*), but those of Engelmann Spruce are not.

Sitka spruce, otherwise called tideland spruce, is the only spruce in region 6. As mentioned in Chapter 2, unlike all other spruces, whose leaves are square in cross section, it is the only spruce with flat leaves. Therefore it could be confused with Douglas-fir or one of the two species of true firs which often grow with it in the West Coast rain forests.

To sort out the four species, note first that the leaves lie flat on either side of a fir twig (Figure 36 shows a twig of Grand Fir as an example), whereas in Sitka Spruce and Douglas-fir the leaves spread out all around the twigs in bottle-brush fashion. There are several conspicuous differences between Sitka Spruce and Douglas-fir. The leafless twigs are different: Those of Sitka Spruce are rough with projecting woody leaf bases, whereas those of

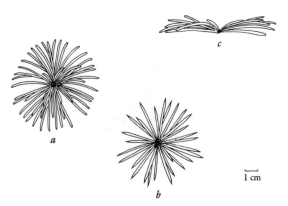

Figure 36. (*a*) Douglas-fir; (*b*) Sitka Spruce; (*c*) Grand Fir.

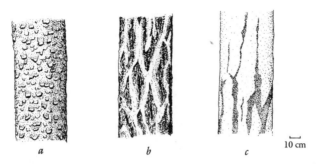

Figure 37. (*a*) Sitka Spruce; (*b*) Douglas-fir; (*c*) Grand Fir.

Douglas-fir are almost smooth (see Figures 6 and 10). The leaves are different: Those of Sitka Spruce are stiff and very sharp-pointed, whereas those of Douglas-fir are too soft to feel prickly. Tall specimens in dense forest can be recognized by their bark. The scaly bark of spruce contrasts strikingly with the deeply furrowed bark of Douglas-fir (Figure 37). (The old bark at the bottom of the trunk in true firs is usually furrowed too, but much less than that of Douglas-fir; see Figure 37*c*).

Finally, of course, there are the cones. The unique trident-shaped bracts of Douglas-fir cones were described in Chapter 2 and are illustrated there and again in Figure 41. The cone of a Sitka Spruce is shown here (Figure 38). It is the most ornamental, and also the largest, of all spruce cones. The individual scales are "waved," making them more or less saddle-shaped, and the cone as a whole has a frilled appearance.

To conclude this account of the spruces in our area, here is a key to them.

Figure 38. Sitka Spruce.   1 cm

Key to the Spruces

1. Leaves almost flat; near the west coast ....Sitka
1. Leaves square in cross section; not near west coast....................2
  2. East of Lake Superior...3
  2. West of Lake Superior...5
    3. Twigs completely hairless ....White
    3. Twigs fuzzy with short brown hairs (use a lens) 4
      4. Cones long and narrow ....Red
      4. Cones short, almost spherical...............................Black
    5. Leaves short (about 1.2 cm long), twigs hairy ....Black
    5. Leaves longer (about 3 cm)...6
      6. Leaves acid-tasting when chewed...........................Blue
      6. Leaves not acid-tasting...7
        7. Cone-scales smooth-edged, scarcely longer than
          seed wings ....White
        7. Cone-scales rough-edged, much longer than
          seed wings..Engelmann

THE HEMLOCKS

There are three hemlocks in our area. They are named, with commendable straightforwardness, Eastern Hemlock, Western Hemlock, and Mountain Hemlock; or, in Latin, *Tsuga canadensis*, *T. heterophylla*, and *T. mertensiana*, respectively. The only source of confusion is the name hemlock itself. Hemlock trees are unrelated to the hemlock that poisoned Socrates. That hemlock (*Conium maculatum*), sometimes called "poison hemlock" to distinguish it from the tree, is a coarse biennial herb of the Parsley family (Umbelliferae); it is said that early European settlers in North America gave the tree the same name as the poisonous weed that they were familiar with because the odor of the tree's leaves, when crushed, recalled the odor of poison hemlock.[2]

Our three hemlock trees are disposed (like our three larches) as follows. There is a single eastern species, of medium size, namely, Eastern Hemlock. There are two western species, Western Hemlock and Mountain Hemlock; the first of these two grows at low elevations in the western rain forests and often attains enormous size; the second is confined to high elevations and is the smallest of the three species.

They are easy to tell apart. The geographic range of the eastern species nowhere approaches that of the two western species. Its western limit is east of Lake Winnipeg and the Red River; hence any wild hemlock found east of these landmarks is certainly an Eastern Hemlock.

The two western species have much the same geographic range and even

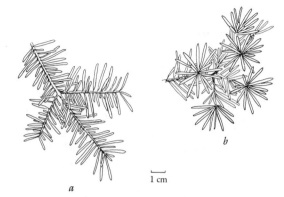

Figure 39. (a) Western Hemlock;
(b) Mountain Hemlock.

though one is a lowland and the other a subalpine species, their elevational ranges overlap. Therefore they are often found growing together, in the broad boundary zone on the mountain slopes that could equally well be called the lowest part of the subalpine forest or the highest part of the mountain-valley forest. However, it is impossible to confuse them as the arrangement of their leaves is quite different.

As shown in Figure 39, the leaves of Western Hemlock are two-ranked; that is, they lie horizontally on either side of the twig. The leaves of Mountain Hemlock, in contrast, tend to form neat, decorative circular rosettes. Another contrast between the two species is that Mountain Hemlock has much thicker leaves than Western Hemlock; this fact is not much use as a diagnostic character unless trees of the two species are growing side by side.

Hemlock cones are not especially noteworthy. Those of the two lowland species are small, only about 2 cm (¾ in.) long. Mountain Hemlock cones are surprisingly big in comparison, up to 4 times as long. As with the larches, the smallest species, growing in subalpine habitats, has the largest cones. Presumably this is merely a coincidence; there is no reason to suppose that large cones are better adapted than small ones to the high mountain environment.

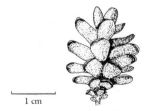

Figure 40. Western Hemlock.

THE DOUGLAS-FIRS

There is only one species of Douglas-fir in our area, *Pseudotsuga men-ziesii*. It is not found east of the Rockies, and where it grows it is unmistakable. It was described in Chapter 2 and all that is needed here is a reminder of its salient characteristics. Its cones, with their three-pronged bracts, are unique. The leaves are flat and the twigs smooth. And the buds are pointed, and papery (as opposed to sticky) to the touch.

Douglas-firs occur in two forms (or varieties or subspecies or geographic races) a coast form and an interior form. In this variability they resemble Lodgepole Pine, which has a short, contorted "shore" form on the West Coast and a tall, straight, "ordinary" form in the mountains of the western interior. But in the case of the Douglas-firs it is the small form that grows in the interior and the large form at the coast. Coast Douglas-firs are enormous trees, up to 100 m (330 ft) tall and 4 m (15 ft) in diameter; indeed, they are the largest trees in our area.

Their cones, too, are larger than those of interior Douglas-firs; they also differ in that, in the coast form, the trident-shaped bracts tend to point forward and lie flat, whereas in the interior form the bracts are often bent back.

The two forms of Douglas-fir, like the two forms of Lodgepole Pine, do not differ from each other enough to be treated as separate species. But given enough time (hundreds of thousands of years? millions of years?), they *could* evolve into separate species. For this to happen, it would not be necessary for all interbreeding to be prevented. The two forms could, over many generations, diverge from each other in their characteristics, each becoming better adapted to its particular environment (very wet at the coast, drier in the interior), in spite of the persistence of a small amount of interbreeding in a narrow zone where their geographic ranges overlap.

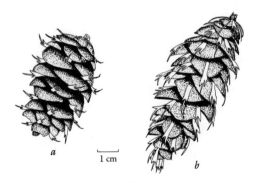

a

1 cm

b

Figure 41. Douglas-fir cones. (*a*) Interior form; (*b*) coast form.

THE FIRS

One of the most noticeable characteristics of the firs (only noted in passing in Chapter 2 because it is not obvious at all seasons) is that their cones sit upright on the branches. Furthermore, a fir cone does not fall in one piece when it is mature; instead, the scales, with their seeds, fall off individually, leaving the bare cone axis still on the tree. Hiking through a fir forest in late summer, we often find the trail carpeted with shed fir cone scales and seeds. This is not, as is often thought, the result of careless foraging by uncountable hordes of squirrels; it comes from the natural disintegration of the cones on the overhanging fir branches.

The cones of fir are usually confined to the topmost branches and are therefore difficult to see if the tree is tall. Look for them on young trees; they are oval to oblong, smooth, plump, and purple when ripe.

There are two wide-ranging fir species in our area, and three with more restricted ranges. First, consider the two widespread species: Balsam Fir (*Abies balsamea*) and Alpine Fir (*Abies lasiocarpa*). The dash-dot line in the map below is a "separation line." Alpine Fir is not found east of it nor Balsam Fir west of it; but note that this does *not* mean that Balsam Fir is found *everywhere* to the east of it and Alpine Fir everywhere to the west of it. The cross-hatched region shows where the ranges of the two species overlap; there, they grow together and hybridize.

The Balsam/Alpine pair of fir species resembles the Jack/Lodgepole pair of pine species, and the White/Engelmann pair of spruce species. In each

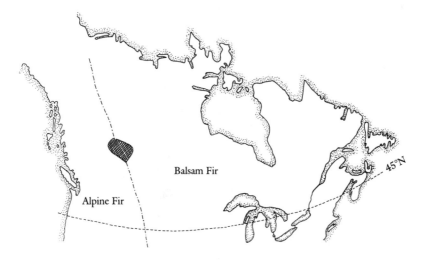

Figure 42. Boundary between the ranges of Balsam Fir and Alpine Fir.

of these three species-pairs, the first-named is a tree of the northwoods (or, more formally, of the boreal forest) and the second a tree of the western mountains. Moreover, the members of each pair are very similar to each other, are closely related, and are able to hybridize.

Balsam Fir and Alpine Fir are so similar to each other that some botanists treat them as a single species, classifying Alpine Fir as a subspecies (or geographic race) of Balsam Fir. The two species (or one, if you prefer) are trees of absolutely unmistakable outline; their tops form tall, smooth, tapering, narrow spires.

No other conifer is like this. Therefore the two species (for we shall here treat them as two) are easy to recognize outside the region where their ranges overlap. Within the region, the two are difficult to tell apart. The clearest difference between them is in the upper surfaces of their leaves, which are a shiny dark green in Balsam Fir, and a dull bluish-grayish green, speckled with tiny white dots in Alpine Fir. The dots (visible with a hand lens), called *stomata*, are explained in Chapter 5.

The two species have very similar cones. A cone of the common or bractless form of Balsam Fir is shown in Figure 44*a*. Strictly speaking, it is not bractless; rather, the bracts are shorter than the scales and are therefore concealed. A variety in which the bracts are longer than the scales is shown in Figure 44*b*. This bracted variety, known as Bracted Balsam Fir, occurs only in the eastern part of the range of Balsam Fir and becomes commoner the farther east we go. In Newfoundland, Bracted Balsam Firs outnumber the bractless form. The three other fir species in our area are not found east of the continental divide. They are Grand Fir (*Abies grandis*); Amabilis Fir

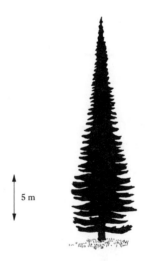

5 m

Figure 43. Alpine Fir.

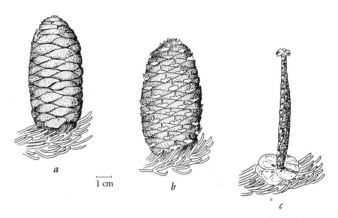

Figure 44. Balsam Fir. (*a*) Bractless cone; (*b*) bracted cone; (*c*) cone axis after the fall of most of the scales.

(*Abies amabilis*), also known as Pacific Silver Fir; and Noble Fir (*Abies procera*).

Noble Fir has the smallest range within our area. It grows only at middle elevations in the forests of the Cascade Mountains south of the latitude of Wenatchee, Washington. It has very distinctive cones. They are large, up to 15 cm (6 in.) long, and have big, protruding bracts that are bent back so that they cover, and almost hide, the scales.

Grand Fir and Amabilis Fir are both fairly common in the Pacific Northwest and adjacent British Columbia. Amabilis is confined to the coastal mountains and is found as far north as the Alaska panhandle. Grand Fir grows in the coastal lowlands, and also in the valleys of both the coastal and interior mountains, but not north of the northern tip of Vancouver Island. There is nothing especially noteworthy about the cones of either

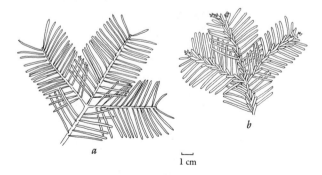

Figure 45. (*a*) Grand Fir; (*b*) Amabilis Fir.

species. They do not have protruding bracts. The best way to distinguish the two species from each other is to notice the arrangement of their leaves.

On a twig of Grand Fir all the leaves are the same length, aligned in two ranks on either side of the twig and not concealing it. By contrast, a twig of Amabilis Fir has leaves of two lengths; there are long leaves in a two-ranked pattern, and also short leaves lying flat on the top of the twig, pointing forward, and overlapping one another like shingles on a roof.

The Arborvitaes

The two arborvitaes in our area (they are the only two in North America) are remarkably different from each other as seen from a distance. They are Eastern Arborvitae (*Thuja occidentalis*), often called "eastern white cedar," a very small tree found only to the east of the prairies; and Giant Arborvitae (*Thuja plicata*), often called "western red cedar," a forest giant of the western rain forests from the Rocky Mountains to the Pacific coast. The size contrast is clear from the drawing. The eastern species is often incredibly neat and symmetrical; it is hard to believe that the shape is natural and not the result of a topiarist's careful clipping and shaping. The western species, Giant Arborvitae, is notable for its size and for the fact that it is one of the few temperate zone trees that frequently develops a buttressed (fluted) trunk in old age. This form of growth is much commoner in tropical trees. Young Giant Arborvitaes have curled-over tips like those of Western Hemlock, but there is no possibility of confusing the two

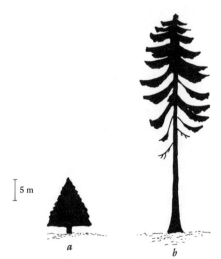

5 m

a

b

Figure 46. (*a*) Eastern Arborvitae; (*b*) Giant Arborvitae.

Figure 47. Giant Arborvitae, but-tressed trunk.

20 cm

genera since the hemlocks have needle-leaves and the arborvitaes scale-leaves (see Figure 13).

The leaves, cones, and bark of the two arborvitaes are very similar. Anyone familiar with the small, eastern species will instantly recognize the giant, western species as an arborvitae and vice versa, in spite of the tremendous size difference.

The cones are small and woody, with only a few scales. In good cone years they grow as continuous mats on the upper surfaces of the frondlike branches. Each seed has a pair of wings, one on each side. In this character they differ from the pines, larches, spruces, hemlocks, Douglas-firs, and firs, all of which (except for the stone pines, whose seeds are wingless) have seeds with a single wing.

The fact that the arborvitaes are known in the lumber trade as cedars is no reason why naturalists should follow suit. It has already been noted

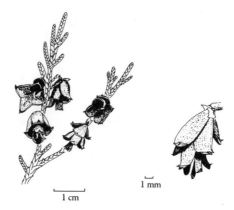

1 mm

1 cm

Figure 48. Giant Arborvitae cones.

1 mm   *Figure 49*. Giant Arborvitae seeds.

that, in the trade, Giant Arborvitae is known as western red cedar and Eastern Arborvitae as eastern white cedar (or sometimes, to add to the confusion, as northern white cedar); in the East the adjective red is transferred to a tree species of juniper known in the industry as eastern red cedar. It is easy to avoid the muddle by using the name Arborvitae, which has an interesting origin.[3] It is a latinized form of *l'arbre de vie*, the tree of life. This name was bestowed on the tree in the early 16th century by King Francis I of France, when the French explorers led by Jacques Cartier were cured of scurvy by a tea brewed from its foliage, which the Indians of the St. Lawrence valley gave them.

FALSE-CYPRESS

There is only one false-cypress in our area, the Nootka False-cypress, *Chamaecyparis nootkatensis*. Like the arborvitaes, the tree has a generous supply of misleading trade names, for example yellow cypress (or cedar) and Alaska cedar (or cypress). False-cypress is the most botanically appropriate English name; the trees are so similar to true cypresses (genus *Cupressus*) that some botanists would lump them all together in the same genus.

False-cypress (we can drop the adjective Nootka as no other species concerns us here) occurs only in the rain forests of the West Coast, and there the only tree for which it could possibly be mistaken is Giant Arborvitae as both species have scale-leaves and twigs that grow as flat sprays. But as shown in Figure 14, the twigs are quite different on close inspection. The bark, too, is conspicuously different in the two trees (Figure 15); to put it succinctly, the bark is stringy in Giant Arborvitae and

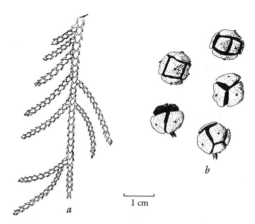

Figure 50. False-cypress. (*a*) Branchlet; (*b*) cones.

1 cm

flaky in False-cypress. The base of the trunk is commonly buttressed, again like that of Giant Arborvitae. The cones are surprisingly small for so large a tree. They are little dark-brown spheres, about 1 cm (less than ½ in.) in diameter. Fallen cones beneath a tree are likely to be dismissed as rabbit scat if they are noticed at all. They take 2 years to mature.

THE JUNIPERS

We have five junipers to consider: two shrubs and three trees.

The two shrubs are Common Juniper (*Juniperus communis*) and Creeping Juniper (*Juniperus horizontalis*). Both have enormous geographic ranges and can be found throughout almost all of our area in a wide variety of (usually dry) habitats. They span the width of the continent from east to west. Common Juniper, but not Creeping Juniper, occurs as far north as the Arctic Circle and beyond.

The ranges of the three tree species are not nearly so extensive. The eastern tree is Eastern Juniper (*Juniperus virginiana*), which is known in the lumber trade as eastern red cedar. Within our area, it occurs only in the Ottawa River valley and along the north shore of Lake Huron, but it is a common tree farther south, in the eastern United States. The remaining two species are western. They are Rocky Mountain Juniper (*Juniperus scopulorum*), which occurs from the Rockies to the Pacific, with a northern limit at about the latitude of Prince Rupert, British Columbia. And Western Juniper (*Juniperus occidentalis*), which is found in our area only at inland localities in southern Washington, Oregon, and adjacent Idaho.

The species are easy to distinguish except where Rocky Mountain Juniper and Western Juniper are both likely to be found.

First, the shrubs: Common Juniper is the only juniper whose leaves are always needlelike—or more precisely, awl-like. Although usually a shrub, it may grow to the stature of a small tree.

Creeping Juniper is always low-growing; hence the name *horizontalis*. Its branches trail over the ground forming a dense, wide-spreading mat. The leaves on the older parts of the branches are scalelike, but the young twigs bear so-called juvenile foliage; this consists of diverging awl-like leaves. Therefore there are two kinds of twigs (Figure 51) and they can be found on the same branch.

Next, the tree junipers. All of these trees, like Creeping Juniper, have both scalelike and juvenile, awl-like leaves. Eastern Juniper is the only tree juniper in the East; it is not commonly a large tree, but in the most hospitable environments it may grow as tall as 10 m (35 ft). Rocky Mountain Juniper and Western Juniper can be recognized, where their ranges overlap, by examining their scalelike leaves with a very strong hand lens; the leaf edge in Western Juniper has a fringe of minute teeth, but in Rocky Mountain Juniper the leaf edge is completely smooth. Western Juniper is much the larger tree of the two; it may be as much as 15 m (50 ft) tall. Rocky Mountain Juniper rarely exceeds 10 m (35 ft) and may be no larger than a shrub if conditions are adverse.

In all the junipers the seed-bearing and pollen-bearing cones are on separate plants (except in the occasional hermaphrodite freak). Each seed-bearing cone consists of a few fleshy scales fused together to form a small blue "berry" with a waxy whitish bloom. That these "berries" are really cones becomes obvious on inspecting their outer ends, where protruding bumps (the tips of the cone-scales) can be seen (see Figure 52). In Eastern and Creeping junipers the cones mature in 1 year. In the other species (Common, Western, and Rocky Mountain junipers), the cones require 2 years to mature, and therefore "berries" of two different ages can usually

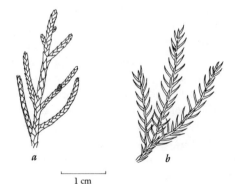

a   b

1 cm

Figure 51. Creeping Juniper.
(*a*) Scale-leaves; (*b*) juvenile leaves.

be found on the same (female) plant: In the first year they are small and green, and in the second year larger and blue. Embedded in the "berries" are from one to five seeds.

THE YEWS

On the subject of the yews there is little to add to what was said in Chapter 2. American Yew, *Taxus canadensis*, is a straggling shrub not found west of Lake Winnipeg and the Red River. Pacific Yew, *Taxus brevifolia*, is a small tree (sometimes a large shrub) not found east of the Rockies. Thus there is no danger of confusing the two species, and the fact that they are yews is obvious from the "berries," provided a female plant can be found. How to recognize a yew by its leaves, as must be done with male plants, was explained in Chapter 2.

A yew "berry" is entirely different from a juniper "berry." It is not, as is a juniper "berry," a fleshy cone. The soft, red, pulpy layer that partly encloses the seed of a yew, and which is known as an *aril*, is quite unrelated to a cone-scale. In the yews, a single seed partly covered by an aril takes the place of a seed-bearing cone.

THE FAMILIES OF CONIFERS

Chapter 2 described the ten different genera of conifers in our area and this chapter has thus far dealt with the species (37 altogether) belonging to these genera. That is, taking the classification into genera as a starting point, we have proceeded to a finer classification, a classification into species. We now reverse the focus and consider the larger groupings, known as *families*, thus proceeding to a coarser classification.

The 10 genera we have considered fall naturally into three families.

The first of these is the Yew family, Taxaceae, with *Taxus*, the yews, as its only genus in our area. The yews are entirely unlike all our other genera in having a single aril-covered seed in place of a many-seeded cone with cone-scales.

Our remaining nine genera fall into two distinct groups. The first consists of the pines, larches, spruces, hemlocks, Douglas-firs, and firs; these belong to the Pine family, Pinaceae. The second consists of the arborvitaes, false-cypresses, and junipers, all belonging to the Cypress family, Cupressaceae.

These two families differ from each other both in foliage and in the structure of their cones. The foliage difference would be clearcut if it were not for the anomalous foliage of Common Juniper. If we exclude Com-

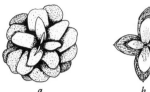

a   *b*   *c*

Figure 52. Seed-bearing cones. (*a*) Hemlock (Pinaceae); (*b*) Arborvitae (Cupressaceae); (*c*) Juniper (Cupressaceae).

mon Juniper, it is true to say that all members of the Pine family have
needlelike leaves and all members of the Cypress family have scalelike
leaves.

The difference between the seed-bearing cones is as follows. In all
members of the Pine family the cones are woody, and have numerous
spirally arranged scales. In the Cypress family, the seed-bearing cones may
be woody (arborvitaes, false-cypresses) or fleshy (junipers). The scales are
far less numerous (sometimes there are only two) and they are arranged
either in opposite pairs, as in arborvitae cones, or in threes, as in juniper
"berries" (Figure 52). (The drawings are not to the same scale.)

The Geographic Ranges of the Northern Conifers

The species described in this chapter are all to be found somewhere in
our area, but none of them is found *only* in our area. All of them grow
south of 45°N as well. Maps of their geographic ranges are given in some
field guides.[4] The most convenient way to summarize the tremendously
detailed information contained in range maps is to sort the species in this
chapter into three geographic groups. (The method is crude and inexact,
as each species has its own unique geographic range that couldn't possibly
be described in one short sentence. But the grouping is useful as long as we
remember its shortcomings.) To group 1 belong those species with southern limits only a short distance to the south of our 45°N boundary; these
are the truly northern species. Group 2 consists of eastern species whose
ranges extend southward into the Appalachian Mountains. Group 3 consists of western species whose ranges extend southward in one or more of
the mountain ranges of western North America. In general, species in
groups 2 and 3 are found at progressively higher elevations in the mountains the farther south you go.

Group 1 contains only 4 species. They are Jack Pine, Western Larch,
Alpine Larch, and White Spruce.

There are 11 species in group 2. Of these, 5 are found only as far south as the latitude of Virginia: Red Pine, Eastern Larch, Black Spruce, Balsam Fir, and American Yew. Two—Red Spruce and Eastern Arborvitae—range south to North Carolina and Tennessee. And 4—Eastern White Pine, Pitch Pine, Eastern Hemlock, and Eastern Juniper—grow as far south as Georgia.

There are 20 species in group 3. Eight of them reach their southern limit (at least as common species) no farther south than the latitude of northern California: Sitka Spruce (which is confined to the coast); Western and Mountain Hemlocks; Grand, Amabilis, and Noble firs; and Giant Arborvitae and Nootka False-cypress. A further 9 species range somewhat farther south, though not into Mexico: Western White, Whitebark and Limber pines, Engelmann and Blue spruces, Alpine Fir, Western and Rocky Mountain junipers, and Pacific Yew. The remaining 3 species in this group have ranges extending into Mexico: Lodgepole and Ponderosa pines and Douglas-fir.

Two species have ranges so large that they cannot be pigeon-holed into any of the three groups. They are Common Juniper, which grows everywhere in our area except in the Far North, and also south of our area both in the Appalachians and throughout the mountains of the western United States; and Creeping Juniper, which also ranges from east to west across the continent, but southward only as far as Colorado, Iowa, and New York.

NOTES

1. R. Daubenmire, "Taxonomic and Ecologic Relationships between *Picea glauca* and *Picea engelmannii*," in *Canadian Journal of Botany*, vol. 52, pp. 1546–60, 1974.

2. R. C. Hosie, *Native Trees of Canada*, 7th ed. (Ottawa: Canadian Forest Service, 1969).

3. W. M. Harlow and E. S. Harrar, *Textbook of Dendrology*, 5th ed. (New York: McGraw-Hill Book Company, 1968).

4. C. F. Brockman, *Trees of North America* (New York: Golden Press, 1968).

Chapter 4

The Reproduction of Conifers

POLLEN CONES AND POLLEN

All the cones described and pictured in Chapter 3 were seed cones. Every species of conifer bears pollen cones as well. They are smaller than the seed cones and do not persist for nearly so long. They shrivel and dry up as soon as the pollen has been shed in spring. Although most of them soon fall off, dried-up dark-brown pollen cones can still be found on the twigs of pine trees right into late fall or winter; they fall off at a touch. In the great majority of conifer species seed cones and pollen cones grow on the same tree. In the yews and the junipers, however, there are separate male and female plants: The male plants bear only pollen cones and the female plants only seed cones. (This statement is not altogether free of exceptions; trees of Western Juniper are occasionally found that bear both types of cone.)

Whatever the species, a pollen cone always consists of an axis with stamens projecting all around it. The stamens are small scalelike organs with pollen sacs attached to them. In species belonging to the Pine family, stamens are always numerous, and they are arranged in a spiral around the axis. Two examples are shown in Figure 54. Notice that the drawings are on different size scales. The pollen cones of Ponderosa Pine are very large, as pollen cones go, whereas those of Douglas-fir are much smaller and more typical.

In the Cypress and Yew families the pollen cones have far fewer stamens, seldom more than a dozen. The stamens are umbrella-shaped, with the

Figure 53. Jack Pine pollen cones.   1 cm

pollen sacs on the inward-facing, protected surfaces of the "umbrellas." The next pair of drawings (Figure 55) show typical examples. Figure 55*a* is a pollen cone from a Western Yew after the pollen has been shed and the "umbrellas" have begun to wilt. Figure 55*b* is a pollen cone of Rocky Mountain Juniper with the pollen sacs showing in the gaps among the "umbrellas." Notice that the pollen cones of the yews, like those of other conifers, have a number of stamens. It is only the unique structure of their berries, each with a single seed, that sets the yews apart in a separate family of conifers.

The pollen of conifers is familiar to anybody who has brushed against a cone-laden branch in spring, releasing dense clouds of yellow pollen. The quantity produced is enormous. It is borne on the wind to female cones

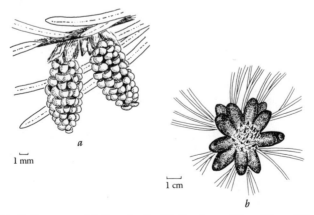

Figure 54. Pollen cones. (*a*) Douglas-fir; (*b*) Ponderosa Pine. (Note different size scales.)

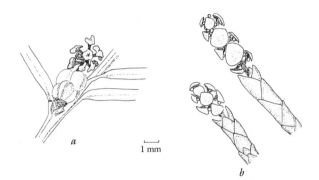

Figure 55. Pollen cones. (*a*) Western Yew; (*b*) Rocky Mountain Juniper.

(seed cones) awaiting fertilization. It blows everywhere and often collects as a yellow film wherever there is a calm water surface—on rain puddles, ponds, and the sheltered backwaters of lakes.

Most people know pollen only as "dust," in air-borne clouds or water-borne films. The individual pollen grains are of much more than passing interest, however. A pollen specialist (a *palynologist*) can identify the species of tree from which a given sample of pollen came by examining the shape and texture of the grains.

Thus the spruces, firs, pines, and Mountain Hemlock (but not the other hemlocks) have grains with a pair of "wings," or bladders, attached. The bladders are often described as wings, but the term is not descriptive unless we visualize inflated water wings rather than, for example, the flat wings of pine seeds. In all species that have them except the spruces, each bladder is constricted at the top as though it were held in by a partly tightened drawstring. There are no such constrictions at the tops of the bladders in spruce grains. The contrast is shown in Figure 56. Compare the bladders on the grains of Red Spruce and Red Pine.

Some conifers, for example, the larches, Douglas-fir, and Eastern and Western Hemlock, have bowl-shaped grains. The surface texture of a grain, as well as its shape, is one of its distinctive characteristics. In Eastern Hemlock, for example, the surface of the grain is minutely knobbly, as shown in the drawing. Douglas-fir grains, which are also bowl-shaped, have a much smoother texture. The grains are tiny, of course. The scale mark in the drawing gives an idea of their size. It is 10 μm long, and 1 μm (micrometer) is one-millionth of a meter—equivalently, one-thousandth of a millimeter (four one-hundred thousandths of an inch). The Red Spruce grain is almost 100 μm, that is, one-tenth of a millimeter, in length. Its bladders would just be visible under a dissecting microscope used at highest power. To examine pollen grains in all their minute details, palynologists use scanning electron microscopes.[1]

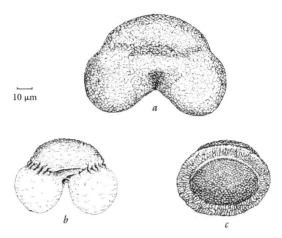

10 μm

Figure 56. Pollen grains. (*a*) Red Spruce; (*b*) Red Pine; (*c*) Eastern Hemlock.

The study of pollen grains is a fascinating subject in its own right, and in addition, it has two important applications. First, a knowledge of what kinds of pollen and how much of each kind is in the air at different times and in different places is of great concern to hay fever sufferers. The second application of palynological knowledge is the study of ancient climates and ancient vegetation. Pollen is produced in enormous quantities every year wherever wind-pollinated plants grow, and it settles everywhere. The grains that alight on the waters of lakes and ponds may form a temporary surface film, but they eventually sink to the bottom and become incorporated in the accumulating sediments. The grains are almost indestructible, and they form so-called microfossils, which persist unchanged, in recognizable form, for millions of years. As a result, it is possible to infer how the vegetation of a region, and hence the regional climate, has varied over past centuries and millennia. The history of vegetation, as revealed by fossil pollen, is a full-fledged science.

POLLINATION

Pollination is the transfer of pollen grains from a pollen cone, in which the grains were formed, to a seed cone, where they will fertilize the ovules (an ovule is a seed before it is fertilized). Everybody is familiar with the pollination of plants with showy flowers, by insects such as bees. When a bee that has become dusted with pollen from one flower visits a second flower, some of the pollen grains it is carrying become stuck to a sticky receptive surface in the second flower, and fertilization proceeds. In plants with inconspicuous, unscented flowers, such as grasses, pollen is blown by the wind from one flower to another instead of being carried by insects.

But the result is the same; the pollen sticks to a receptive surface on a part of the receiving flower that *contains* the ovules awaiting fertilization.

The important point to note is that in flowering plants, as opposed to conifers, the pollen grains do not land directly on the ovules but on closed organs (ovaries) containing them. In the conifers, the wind-borne pollen lands directly on the ovules, which are not enclosed in ovaries, and for this reason the conifers are known as *gymnosperms*, from the Greek words *gymnos* (naked) and *sperma* (seed).

Naturalists are often puzzled to read that the seeds in, for example, a young pine cone are "naked." The reverse seems to be true; few things are more tightly sealed than a hard, green pine cone before it dries, opens, and sheds its seeds; and it is difficult to think of anything less naked than the seeds inside. The puzzle arises because the cones are open only when they are newly formed and before they are pollinated. At this stage they are still very small and immature, and are seldom noticed. They continue to grow after pollination and by the time they reach maturity, after a year's growth, they are large, hard, green, and very tightly closed (see Figure 16*b*).

The way pollination takes place is this: In spring, before the pollen cones have shed their pollen, the season's new, immature seed cones are small and soft. (Some botanists describe such unpollinated cones as "flowers.") The scales are slightly separated and there is a small amount of fluid, secreted by the plant itself, in the narrow crevices between them. When the pollen cones discharge their clouds of pollen, some of it sifts down between the scales of nearby seed cones and comes to rest floating on the fluid. The fluid is then absorbed by the cone, and the pollen trapped on its surface is drawn down into the crevice until it comes to rest on one of the two ovules attached to the bottom of each cone-scale. The pollen grains come into direct contact with the ovules, which are not encased in an ovary.

As soon as most of the ovules in a cone have been pollinated in this way, the cone-scales grow thicker until they are firmly pressed together. No crevices remain into which pollen could possibly drift. The closed cone gradually grows to full size, and fertilization of the ovules proceeds inside.

A Contrast between Seed Cones and Pollen Cones

An interesting point to notice about a conifer's cones is that pollen cones and seed cones are not equivalent to each other in the anatomical sense. The giveaway is the arrangement of the bracts in the two kinds of cones.

Bracts are small, flat, modified leaves, and in seed cones there is a bract behind every scale. In some species the bracts are longer than the scales and are easily seen without breaking the cone apart. In our area, this is true of Douglas-fir, Western and Alpine Larch, Noble Fir, and the bracted variety of Balsam Fir. (See Figures 31*b*, 41, and 44*b* for examples.) In all other species the bracts are small and hidden by the scales; to see them, you must break open a cone so that the backs of some of the scales are exposed. A couple of examples are shown in Figure 57*a* and *b*. In the cones of Arborvitae and False-cypress each bract has become fused with its scale, so that examination with a microscope is required to show that what looks like a scale is really a combination of scale and bract. But no matter what the species, in seed cones each scale is always backed by a bract.

In the pollen cones, on the other hand, there is a bract at the base of each whole cone, not at the base of each stamen (see Figure 58). This contrast leads to a surprising conclusion. If you compare the flowers of the flowering plants with the cones of conifers, it becomes clear that whereas a pollen cone is the equivalent of a single flower, a seed cone is the equivalent of a cluster of flowers (an *inflorescence*), each scale representing one complete flower.

In such a comparison *within* the conifers, a single scale from a seed cone is equivalent to a whole pollen cone; or, what comes to the same thing, a

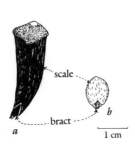

Figure 57. Seed-cone scales (outer side). (*a*) Eastern White Pine; (*b*) Western Hemlock.

Figure 58. Cluster of pollen cones of Eastern White Pine.

single seed cone is equivalent, anatomically speaking, to a whole cluster of pollen cones.

NOTES

1. I. J. Bassett, C. W. Crompton, and J. A. Parmelee, *An Atlas of Airborne Pollen Grains and Fungus Spores of Canada* (Ottawa: Research Branch, Canada Department of Agriculture, Monograph No. 18, 1978).

Chapter 5

The Life and Growth of a Conifer

There are two fundamental differences between the conifers and the flowering trees (all our broad-leaved trees have flowers, even though they are often inconspicuous). The first difference, as already described in Chapter 4, is in the reproductive arrangements. In the conifers, the seeds are borne in cones and are "naked," that is, not enclosed in an ovary. In the flowering trees the seeds are always contained inside a closed ovary and are therefore not naked.

The second difference is in the "pipes" that carry the sap, from the roots to the trunk, from the trunk to the branches, and finally to the ultimate twigs and the leaves. In the conifers there are no real pipes; there are merely chains of long, narrow cells, known as *tracheids*, linked to each other through small, porous membranes. In the flowering trees, by contrast, true conducting pipes carry the sap throughout the plant; they are called *vessels*. A vessel consists of a chain of cylindrical cells lined up end to end; when first formed, these cells have thin end walls that form partitions, but the end walls soon dissolve, leaving the vessel unobstructed.

Figure 59 shows the contrast between tracheids and vessels. Most tracheids have diameters of less than one-twentieth of a millimeter (0.02 in.) while those of vessels are often 20 times as much. A tracheid's length is more than 100 times as great as its diameter, and it is tapered at both ends. In a chain of connected tracheids, each tracheid overlaps those above and below it so that the tapered ends are pressed against each other. In these tapered ends are the holes, covered by porous membranes, through which each tracheid is linked to the next.

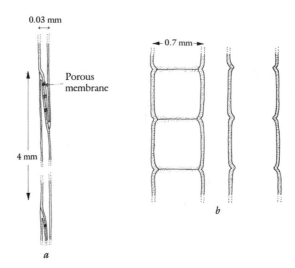

Figure 59. Diagrams of (*a*) tracheids and (*b*) vessels: (left) when first formed; (right) after disintegration of cross walls.

It is this difference between the conifers and the flowering plants—*all* flowering plants, herbs as well as trees—that explains the name *angiosperms* for the flowering plants in contrast to *gymnosperms* for the conifers. The word angiosperm is a combination of the Greek words *angeion* (vessel) and *sperma* (seed), from which you are meant to infer than an angiosperm is a seed plant in which the sap flows through continuous pipes or vessels.

The sap-conducting system is more primitive in the conifers than in the flowering plants. Tracheids function much less efficiently than do vessels as pipes to conduct the sap. The rate of flow of the sap in conifers is about half a meter (20 in.) per hour on average, a snail's pace compared with rates in excess of 20 meters (65 ft) per hour in some oaks. Also, tracheids are less specialized than vessels in that they perform two separate functions: Not only do they carry the sap, they also give wood its mechanical strength.[1]

In the more highly evolved flowering trees, there is a division of labor among the cells in the trunk: The two functions, sap-conduction and mechanical support, are performed by different kinds of cells. Thus the vessels serve only as sap-conductors; they haven't the strength and rigidity required to support a tree. Strength is provided by wood fibers—long, narrow, thick-walled cells whose only function is to provide mechanical support.

The importance of tracheids to coniferous trees should now be obvious. They are found in all parts of the tree, and coniferous wood consists almost entirely of tracheids. They are popularly known as wood fibers, although in botanical parlance only the strengthening cells in the wood of flowering trees, described above, are given the name fibers. As an individual tracheid is too small to examine with a hand lens, it is an inconvenient object for a field naturalist to study. However, the wood of conifers, and other parts of the trees such as bark, roots, and leaves, have many interesting characteristics that can be seen without a microscope. They are the topics of the next sections.

THE WOOD OF CONIFERS

The cut stump of an evergreen, with its concentric annual rings from which the tree's age can be counted, is a familiar sight to everyone. The rings are formed as the tree grows. Just inside the bark, between it and the wood, is a very thin sheet of living cells known as the *cambium*. The cambium forms a sleeve, one cell thick, entirely enclosing the wood. Its cells are too small to be seen without a microscope.

Throughout every year's growing season the cambium cells divide repeatedly. They split in half lengthwise; at each division a wall forms, dividing what was previously a single cell into two "daughter" cells, one toward the outside of the trunk, the other toward the inside. One of these daughters, most often the outer one but occasionally the inner one, becomes a cambium cell itself, and thus replaces its "mother" cell. What happens to the other daughter depends on whether it is the inner or the outer of the two daughter cells. If it is the outer one it goes to form the bark (see the next section). If it is the inner one it goes to form new wood. In doing so it divides lengthwise a few more times and the resulting long, narrow cells become tracheids.

New tracheids are formed throughout the growing season. Those formed in spring are comparatively large in diameter and have thin walls; they form "spring wood," sometimes called "early wood." Those formed later in the season, as fall approaches and growth slows, are much narrower and have thick walls; they form "summer wood," also called "late wood." Because it is composed of thick-walled tracheids, summer wood appears darker than spring wood, hence the annual rings. Annual rings are much more distinct in conifers than in the majority of flowering trees. The reason is that, unlike the tracheids of conifers, the vessels in most of the flowering trees (that is, the broad-leaved trees, or hardwoods) are all roughly the same size; there is not nearly so much contrast between vessels formed in summer and those formed in spring.

Figure 60 shows how a cut conifer stump appears to the unaided eye

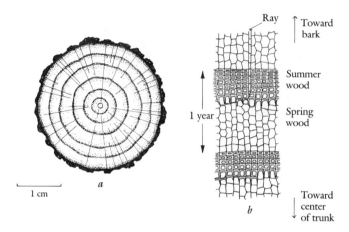

Figure 60. (*a*) Cross section of 7-year-old pine stem; (*b*) micrograph showing the tracheids in cross section.

and, on the right, how a small part of it would appear under the microscope. As the micrograph shows, the tracheids form neat, regular rows. Through every growing season throughout the life of the tree, the dividing cambium adds new tracheids, one at a time, on the outer side of those formed earlier. Consequently the trunk steadily expands as layer upon layer of new wood is formed, summer wood around the outside of spring wood and then spring wood around the outside of summer wood, year after year. As the cylinder of wood grows in girth, the cambium grows to accommodate it, but remains always one cell thick.

The micrograph also shows that almost all the cells of which wood consists are tracheids. But not quite all; there are also *rays*, which are bands of small living cells that serve as connecting links between the tracheids. The way in which ray cells are arranged can be seen in Figure 60, which shows how a ray appears in a cross section of a trunk; it is only one cell wide and extends lengthwise along a radius of the trunk. Figure 61*b* shows how conifer wood appears under the microscope in tangential section (cut as shown in Figure 61*a*). The long, narrow cells are the tracheids; the vertical chains of small cells are the rays seen in end view.

The rays are much bigger and easier to see in the wood of flowering trees (hardwoods) than in that of conifers (softwoods). But if you examine the cross section of a conifer stem that has been cleanly cut and then sanded, the rays can usually be seen with a hand lens as thin, light-colored lines radiating outward. Ray cells have weaker walls than tracheids. When a cut stump dries out, it is the walls of the rays cells that rupture under tension, producing the characteristic radiating cracks in an old stump.

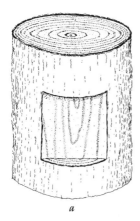 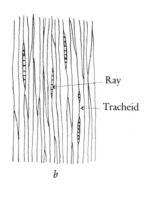

Ray

Tracheid

a

b

Figure 61. (*a*) Tangential section of a pine trunk; (*b*) micrograph showing the tracheids and rays.

Apart from tracheids and rays, the only other structures to be found in conifer wood are resin ducts: hollow channels lined with small living cells that secrete resin. Resin ducts are found in the woods of pine, spruce, larch, and Douglas-fir, but not in the other genera.

The only cells in conifer wood that remain alive for any length of time are the ray cells and the resin-secreting cells lining the resin ducts (in those trees that have resin ducts). The tracheids, which form the great bulk of the wood, die soon after being formed. All functioning tracheids are dead. However, some tracheids are deader than others. The wood in a full-grown conifer trunk is of two kinds: heartwood at the core of the trunk and sapwood around the outside. The tracheids in the heartwood are not merely dead, they are also clogged with resins, gums, and tannins; they cannot conduct sap and the only function remaining to them, albeit a very important one, is as mechanical support for the tree. The tracheids in sapwood, on the other hand, though dead, are free of obstructions and so can conduct sap.

The contrast between heartwood and sapwood explains why a hollow tree is able to survive and flourish. It is only the heartwood that has rotted away; no sap-conducting tissue is lost. Indeed the living part of a tree trunk is really a tall, hollow, tapering cone, perched like a dunce's cap on a solid conical core of heartwood (see Figure 62).

In some conifers—for example, some of the pines and also Giant Arbor-vitae and Douglas-fir—there is a strong color contrast between heartwood and sapwood, as can be seen by looking at a cut stump. In these species, the heartwood is usually brown or reddish brown and the sapwood is white, cream-colored, or light yellow. In firs and spruces, on the other hand, the

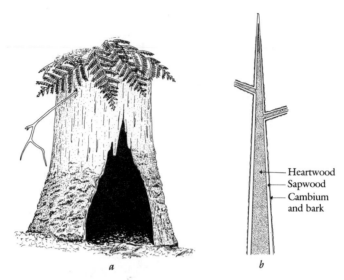

Heartwood
Sapwood
Cambium
and bark

a   *b*

Figure 62. (*a*) A living hollow tree (only the heartwood has rotted); (*b*) the layers of a tree.

color difference between the two kinds of wood is barely noticeable. The material that plugs the tracheids gives heartwood its color, which varies from one species to another.

The thickness of sapwood varies considerably from tree to tree within one species. Sapwood becomes transformed into heartwood sooner in fast-growing trees than in slow-growing ones. Nevertheless, the sapwood layer is usually thicker in the fast-growers because their annual rings are wider.

Cut stumps have still more to offer the naturalist. There is a very interesting difference between the conifers and the hardwoods in the way leaning trees grow. When a growing column of wood is not vertical (and this is true of branches, of course, as well as of leaning trunks), the lopsidedness causes formation of what is called *reaction wood*, marked by wider-than-normal annual rings. In conifers, the reaction wood forms on the lower side, where the stresses acting on the wood tend to compress it; it is therefore called *compression wood* (Figure 63*a*). In the hardwoods, reaction wood forms on the upper side, where the wood is under tension; it is therefore called *tension wood* (Figure 63*b*). Compression wood (the reaction wood of the conifers) is often darker in color than normal wood. Moreover, it has many undesirable characteristics from the wood user's point of view. If used for lumber it is likely to twist and warp during

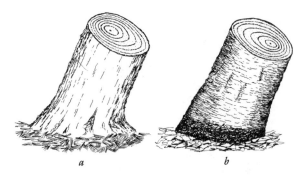

Figure 63. Reaction wood (diagrammatic). (*a*) Conifer (compression wood); (*b*) hardwood (tension wood). The wider rings are the reaction wood.

seasoning; and because it differs chemically from normal wood, it is not much use as pulp wood.[2]

Another property of wood that can be seen without a microscope is its spiral grain, which is present to some extent in nearly all conifers. A straight-grained tree, with its chains of tracheids ranged vertically, is a rarity. Spiral grain is difficult to detect in most living conifers because the grain of the wood is concealed by the bark. The arborvitaes are exceptions; the long fibers that make up their outermost layers of bark often have a spiral pattern themselves, corresponding to the arrangement of the underlying wood cells. Spiral grain is most easily seen in a burned-over forest where the trees have been killed but not consumed, and where the charred bark has fallen away, leaving the dead wood exposed. (The tree in Figure 64 is a Lodgepole Pine in a 12-year-old burn in the Rocky Mountains.) We often find that all, or nearly all, the trees in such a forest have spirals going the same way, either all with a left spiral or all with a right spiral. A right spiral is one in which the grain on the near side goes from lower left to upper right like the ridges on a right-hand screw, and vice versa for a left spiral. Foresters have found that the grain of the wood in a young coniferous tree usually has a left spiral. As the tree grows older, the left spiral steadily diminishes until the grain is straight. The change of direction doesn't stop, however; the twist slowly converts to a right spiral, which increases with age.[3]

Not all trees go through this cycle of changes; there are exceptions. And the ages at which the changes become visible are extremely variable; a tree going through the cycle may be anywhere from 70 to 130 years old before it develops a right-spiral twist. The fact that most trees go through the cycle explains why we often find the majority of trees in a burn twisted in

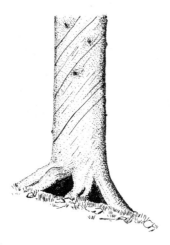

Figure 64. Lodgepole Pine with right-spiral grain (burned tree with no bark).

parallel. Such trees are usually all of the same age—they all began growth together following a much earlier fire—and therefore most of them are at the same point in the "twisting cycle."

The Bark

Tree trunks and branches, although made up chiefly of wood, also have "bark"; this catchall name describes all the tissues, whatever their function, outside the cambium layer. Like wood, bark has two functions. Whereas wood conducts sap upward and provides mechanical support for the tree, bark conducts dissolved food downward and, being waterproof, protects the whole tree against drying out. But whereas cells of only one kind—the tracheids—perform the two functions of wood, different kinds of cells perform the two functions of bark.

Materials have to be conducted both upward and downward in trees, and indeed in all plants. The upward stream consists of sap flowing through the wood; sap is made up of water and dissolved soil minerals absorbed by the roots. In addition, it carries dissolved organic foods (mainly sugars, as maple sap demonstrates) that have been stored at a low level and are needed for growth at a higher level in the tree. The downward stream consists of a rich broth of dissolved organic foods made in the leaves by photosynthesis and required for growth throughout the tree. What is so surprising at first thought is that the downward stream should occupy so little space compared with the upward stream; sapwood forms a much thicker layer than bark.

The explanation is that an enormous quantity of water passes from the soil, up through the trunk and branches, and out to the atmosphere via the leaves; there is no matching downward flow, and thus the volume of liquid going upward greatly exceeds the volume going downward. This tremendous upward flow of water will be discussed again later.

As in the conducting cells in wood, so also in the conducting cells in bark, there is a contrast between the conifers (gymnosperms) and the flowering trees, or hardwoods (angiosperms). In the conifers, these cells, known as *sieve cells*, are long and narrow and are linked together by pores. In the flowering trees, the cells are joined end to end to form *sieve tubes*. Sieve tubes are not completely clear of obstructions, as are the vessels in wood, but they do form much more efficient conducting pipes than the chains of sieve cells in conifers.

New sieve cells are formed in a conifer trunk in the same way as new tracheids, by the division of a cambium cell into an inner and an outer daughter cell. As described in the preceding section, the usual course of events in such a division is for the outer daughter cell to remain as part of the cambium and for the inner daughter cell to become part of the wood. Occasionally, however, at about one division in ten, it is the inner cell that remains as a cambium cell whereas the outer cell, on the bark side of the cambium, develops into something new: It undergoes a few more divisions and the cells produced develop into functioning sieve cells. Unlike tracheids, sieve cells remain alive.

Thus sieve cells are produced *outside* the cylinder of cambium, layer *within* layer, in exactly the same way (though much more slowly) as tracheids are produced *inside* the cambium, layer *upon* layer. Notice the difference. The wood, as it grows, is not under any pressure; the "sleeve" of cambium surrounding it expands to accommodate it. The bark, in contrast, surrounds a steadily expanding column of rigid wood and cannot expand with it. Instead, the outermost layers of conducting tissue (and the tissues outside them, discussed below) are crushed by the pressure and disintegrate. The life of a sieve cell is short, usually less than a year. Thus bark, unlike wood, does not have annual rings.

The protective function of bark is performed by the outermost bark tissues, outside the layers of conducting tissue. The outermost layer consists of cork and is very similar in composition to commercial cork (which comes from the bark of a species of oak). Cork is waterproof, and thus the bark protects a tree from dehydration. It is also a good heat insulator, protecting the living tissues it covers from the injurious effects of abrupt atmospheric temperature changes and also from low-intensity forest fires.

Cork cells are formed throughout the life of a tree by a *cork cambium*, a layer of dividing cells outside the sieve cells. In the firs, the cork cambium forms a continuous sleeve, which, for most of the tree's life, enlarges as the tree grows. Thus firs have smooth bark until they are quite old. In all the other conifers, the cork cambium is itself ruptured by the expansion of everything within it. Because they are being continually torn apart, the outermost tissues are seamed, cracked, and fissured; the outer surface wears away, in discarded flakes and strands. The exact details of this tissue destruction depend on the species of the tree; each species has its own distinctive bark pattern, of scales or plates or ridges, and each pattern results from a mode of break-up of dead tissue peculiar to the species.

THE ROOTS

Because they are subterranean and hard to get at, tree roots do not have much to offer the amateur naturalist. Nevertheless, interesting above-ground evidence can often be found of what is going on below ground. Mushrooms of many kinds are a characteristic part of the ground vegetation in conifer forests, especially in the fall, and a great many of them are closely linked to tree roots in the soil beneath them.

A mushroom is only a temporary above-ground outgrowth from a subterranean fungus whose permanent "body" consists of an extensive network of branching strands ramifying through the soil. It is easy to find a few of these fungus strands (known as *hyphae*) for inspection. They are the fine, white threads that adhere to the undersides of fallen branches and logs lying on the soil. In most forest soils enormous numbers of fungus hyphae grow in all directions. When they come in contact with a soft young tree root, the hyphae grow to surround it, forming a feltlike sheath encasing the root tip. Hyphae also penetrate to the interior of the root and grow in the spaces among the root's cells. What results is a structure consisting of a tree-root-plus-fungus so completely grown together as to constitute a single organ known as a *mycorrhiza*, or fungus-root. The association between fungus and tree is *symbiotic*; that is, it benefits both partners. The fungus gains food (mainly sugars) from the tree root. The tree gains what amounts to a tremendous increase in the absorbing surface of its roots. Roots without fungi absorb soil water through their root hairs, which are too short to reach far into the soil surrounding them. Roots with fungi— that is, mycorrhizae—are able to absorb water and mineral nutrients from a far greater quantity of soil; the fungus hyphae keep on growing, branching repeatedly as they do so, until they have penetrated every pore in a very large soil volume. Once a tree root has formed a partnership with a fungus it usually stops growing root hairs of its own, evidence that the fungus hyphae have taken over the root hairs' absorbing function. Ninety percent or more of all forest trees (hardwoods as well as conifers) have mycor-rhizae; those that haven't, don't thrive. Most of the mushrooms in a conifer forest are the above-ground parts of mycorrhizal fungi.

Mushroom species are numerous; some of them can form mycorrhizae with several different tree species, others with only a few tree species or even only one. Figure 65 shows five of the common mushrooms whose underground hyphae form mycorrhizae with conifers.

Another interesting fact about tree roots is that the roots of neighboring trees easily become grafted to each other if they grow close enough to come into contact. Natural root grafting is common in Eastern White Pine, Red Pine, and Balsam Fir, and probably in other species too. Some-times the roots of a whole stand of trees (of one species) can form an interconnected maze so that the stand functions as a single organism. Dissolved foods flow throughout the common root system from one tree to another, enabling small, overshadowed trees, or cut stumps, to "steal" food from their taller sunlit neighbors. The latter, because they receive plenty of light, can carry on photosynthesis to feed both themselves and the less successful trees that happen to be grafted to them. If the stump of a felled tree in a forest remains alive and fresh for years, it is a sign that its roots are still alive and are grafted to those of the surrounding living trees. Indeed, a single large tree can provide enough food to keep many stumps alive, but they will all quickly die if their provider tree is felled.[4]

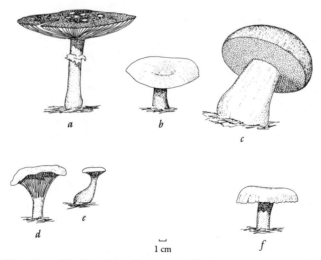

Figure 65. A few of the fungi that form mycorrhizae. (*a*) Fly Agaric; (*b*) Peppery Milkcap; (*c*) Edible Boletus; (*d*) Chantarelle; (*e*) Clubfoot Clitocybe; (*f*) Stinking Russula.

THE LEAVES

The leaves of conifers, as of all plants, are the trees' "food factories." As everyone knows, it is in the leaves that photosynthesis takes place, the formation of sugars from water and carbon dioxide in a reaction powered by sunlight and involving the green chemical chlorophyll. Therefore a tree can grow only when it has an abundance of leaves, and there are usually several millions on a full-grown conifer.

Of the two ingredients needed for the manufacture of sugars, one, carbon dioxide, comes from the air; the other, water, comes up from the soil through the tracheids of roots, trunk, branches, and twigs. For the ingredients to react, they have to come together; and air, bearing its minute proportion of carbon dioxide (about 3 parts in 10,000) has to get inside the leaves. This it does through tiny pores called *stomata* (singular, *stoma*) in the leaf surfaces.

The stomata provide the only route by which air, and its contained carbon dioxide, can reach the internal spaces in the leaves, whence it can be absorbed, through their thin walls, into the cells that contain chlorophyll. If it were not for the stomata, air would be unable to reach the interior of the leaves, because leaves have an impermeable "skin," or *cuticle*. Many needle-leaved conifers also have a layer of wax on top of the cuticle. The wax is the cause of the bluish bloom on such leaves as those of the white pines and the spruces (except Red Spruce), and it makes the leaves unwettable.

The stomata are small oval pores less than one-twentieth of a millimeter long, and they are very numerous. They can be seen with a strong hand lens as tiny white dots ranged in rows along the leaves. The white stripes on hemlock, fir, Douglas-fir, and Common Juniper leaves are bands of closely packed stomata. Each stoma is sunk in a pit below the general leaf surface (see Figure 66), and the opaque white material that makes the visible dot is wax that almost (but not quite) blocks the pore.

A stoma, the gap between two *guard cells*, can open and close. When the guard cells are replete with water and fully distended, they curve and gape apart so that the stoma opens. When the guard cells are not distended with water, they straighten and the stoma closes. Usually the stomata are open only during daylight, when photosynthesis is in progress.

The results of photosynthesis are impressive. On average, one hectare of conifer forest manufactures about 10 metric tons* of plant matter per year. This is dry weight, the weight of the newly formed plant parts after they

---

*One hectare is 10,000 square meters, or the area of a square with sides of 100 meters. It is about 2.47 acres. One metric ton is 1000 kilograms, or about 2200 lb.

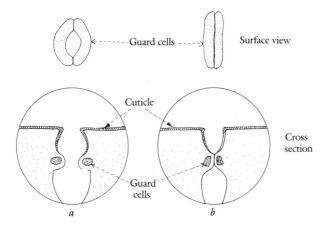

Figure 66. Stoma of a pine leaf (diagrammatic). (*a*) Open; (*b*) closed.

have been dried in an oven to evaporate the contained water. The fresh weight of living material is more than twice as great, 20 to 25 tons at least.[5]

This brings us to the matter of water. In daylight, not only does fresh air charged with carbon dioxide seep into the leaves through the open stomata, water vapor seeps out. The process is called *transpiration*. In forested country a truly prodigious amount of water passes from the soil to the atmosphere via the trees on a sunny day. For example, 1 hectare of Douglas-fir forest can transpire more than 50 tons of water on a single day in summer, a much greater weight than that of the living tissue gained over the same area, as the result of photosynthesis, in a whole growing season. This explains why there is a far greater need for upward-conducting tracheids than for downward-conducting sieve cells in the trunk of a tree, in other words, why there is so much more wood than bark. Far more material goes up than comes down and the surplus—the excess water—is transpired; it evaporates from the thin-walled internal cells in the leaves into the spaces among the cells, and the vapor then diffuses out through the open stomata.

All this water performs a variety of functions. Acting as an ingredient in photosynthesis, in which molecules of water and carbon dioxide unite to form sugar, is only one of the functions. In addition, plain water is an important constituent of all the living parts of a tree and accounts for more than half the tree's total weight; it imparts rigidity (*turgor*) to cells that would otherwise be flaccid. The current of water ascending from the soil via the roots carries dissolved mineral nutrients to the parts of the tree where they are needed. Lastly, evaporation from the leaves has the same effect as sweating; it cools the leaves and prevents their becoming injuriously overheated under a hot sun.

The forces that carry the water through a tree from soil to sky have long been a topic of botanical research. It is by no means obvious what causes sap to ascend. Capillary action—the tendency for water to creep up a very fine tube, which explains why water is absorbed by a sponge—is completely inadequate as an explanation. Although tracheids are narrow, they are not nearly narrow enough to allow the capillary ascent of water to the top of a tall tree. Root pressure, which is exerted when root hairs "suck" in soil water, is probably part of the explanation, but it can be only a small part as it does not provide nearly enough lifting force. The most widely accepted explanation is the "cohesion theory," which is based on the fact that water cannot be stretched; any body of water is held in one piece by an extremely strong force of cohesion. Moreover, water adheres equally strongly to the inner walls of the tracheids. Therefore, provided the "plumbing" of a tree is totally filled with water and free of bubbles from infancy onward, whenever water is removed at the top of the tree—as it is by the growth of new tissue and by transpiration—replacement water is literally dragged up from below. The theory does not require perfectly unbroken, permanently bubble-free columns of sap in every chain of tracheids. Some of the sap columns are likely to rupture, and bubbles to form, when a tree is swayed by the wind. This doesn't matter. For one thing, most trees have far more tracheids than they need; for another, when bubbles are formed they are full of water vapor and don't take long to contract and vanish. The mystery of the ascent of sap seems to be explained. The force that brings it about originates in the leaves.

The remarkable differences between conifer leaves and the broad, thin leaves of the hardwoods are explained by the fact that conifer leaves in our area (except those of the larches) are evergreen. Because the leaves remain on the trees all through the winter, when temperatures are often too low for the roots to function, they are adapted to conserve water. The water in a tree must obviously be conserved at times when it cannot be replenished. A very noticeable adaptation is that the needle-shaped leaves of conifers have a low surface-to-volume ratio; even so, the total area of leaf surface in an entire conifer tree is comparable to that of a hardwood tree of the same size, because conifer leaves are so numerous. Transpiration is slowed, and water conserved, by the thick waterproof cuticles of the leaves and by the fact that the stomata are sunk in pits instead of being flush with the rest of the leaf surface as they are in hardwood leaves. These adaptations also give strength to the leaves, and we can fairly say that conifer leaves are built to last; some persist on a tree for as long as 8 years or even more. Most evergreen leaves have much shorter life-spans, however; the 1-year-old leaves always outnumber the 2-year-old leaves, which in turn outnumber the 3-year-old leaves, and so on. The younger the leaf, once it has grown to full size, the more efficient it is as a photosynthesizer. The same statements apply to the scale-covered green twigs of junipers, arborvitaes, and false-cypresses, in which whole twigs are the units that correspond to the needles of needle-leaved species.

The last point to note—a cheering one to those who live in cold climates—is that conifers manage to carry on photosynthesis on many days in winter. They can continue to do so at temperatures as low as $-7°$ Celsius (20°F) and perhaps even lower.[6]

NOTES

1. P. J. Kramer and T. T. Kozlowski, *Physiology of Trees* (New York: McGraw-Hill Book Company, 1960).

2. E. J. Mullins and T. S. McKnight (eds.), *Canadian Woods: Their Properties and Uses*, 3d ed. (Toronto: University of Toronto Press, 1981).

3. A. F. Noskowiak, "Spiral Grain in Trees: A Review," *in Forest Products Journal*, vol. 13, pp. 266–75, 1963.

4. F. H. Bormann, "Root Grafting and Non-competitive Relations between Trees," *in Tree Growth* (T. T. Kozlowski, ed.) (New York: The Ronald Press, 1962).

5. J. S. Olson, "Productivity of Forest Ecosystems," *in Productivity of World Ecosystems* (Washington, D.C.: National Academy of Sciences, 1975).

6. P. J. Kramer and T. T. Kozlowski, *Physiology of Trees* (New York: McGraw-Hill Book Company, 1960).

Chapter 6

Insect Pests of Conifers

A conifer forest in which none of the trees is diseased or pest-ridden probably doesn't exist. Forest pests and diseases are as much a part of the natural scene as the trees they afflict, and to judge them in human terms as evils is as anthropocentric as believing that the earth is at the center of the universe. Of course, the anthropocentric view is not surprising, especially if trees are grown for profit or, indeed, if wood or its derivatives (principally paper) are used for any purpose whatever. Even so, to treat a diseased or pest-infested tree as imperfect and therefore uninteresting is a great mistake. Observing and studying the pests and diseases of trees during a walk in the woods adds tremendously to our knowledge of natural history. Naturalists who haven't tried it will find a whole new field of interest awaiting them if they do.

For conifers, most of the pests are insects. Some mammals do a certain amount of damage to trees, but with one notable exception (see Chapter 9), their activities are negligible compared with those of insects. The diseases of trees are chiefly fungus diseases. Although a few tree diseases are caused by bacteria and viruses, infection by damaging (*pathogenic*) fungi is the principal cause of disease.

This chapter describes a few of the insects that attack conifers. Although hundreds of pest insects are known, those chosen for description here are noteworthy either because of their seriousness or because of characteristics that are particularly interesting to naturalists. Chapter 7 describes a few of the fungus diseases, chosen for the same reasons.

MOTHS AND BUTTERFLIES

A number of forest pests belong to the insect order Lepidoptera, the moths and butterflies. One of the most famous, and most serious, is the Spruce Budworm.[1] Its Latin name is *Choristoneura fumiferana*. The name budworm is unfortunate as the animal doing the damage, the larva of a small moth, is not a worm but a caterpillar. Worms proper belong to a different phylum of the animal kingdom. Biologically worms differ from caterpillars (the correct name for the larvae of moths and butterflies) as greatly as snails, say, differ from dogs. And no worms will be found on trees; the commonest worms, the dew-worms or earth-worms, live in the soil. Nevertheless, many economically important caterpillars are saddled with the name worm, for example, inchworms, armyworms, cabbage-worms, and cutworms. (The caterpillar in *Alice's Adventures in Wonderland* would have been outraged.)

To return to the Spruce Budworm. It is a serious pest in eastern North America, especially in Quebec, New Brunswick, and Maine. Notwith-standing its name, its chief victim is Balsam Fir; the spruces come second after firs in its list of preferences. The adult moth is a dull gray-brown and has a wing span of less than 2 cm (0.8 in.); an individual moth of the species is therefore inconspicuous. These insects force themselves on our attention only when there are great numbers of them fluttering around in midsummer. The adult female moths lay their eggs on fir or spruce leaves in summer; the caterpillars that hatch spin themselves silk cocoons, in which they safely survive the winter in a state of suspended animation. The following spring they emerge from their cocoons and the damage begins. As the caterpillars grow, they eat everything within reach—the leaves, buds, young twigs, and young cones of the tree they inhabit. In severe epidemics, they can totally defoliate whole forests of fir and spruce. A walk through an infested forest is unpleasant; many of the feeding caterpillars fall from the leaves and, in falling, spin silk threads from which they hang suspended in midair. These threads stick to the passerby, who may soon become covered with them.

When a caterpillar is full grown, it turns into a chrysalis (*pupa*). Some-times, though not always, a caterpillar will build itself a tent of silk threads and leaf fragments to house its chrysalis. (Figure 67 shows a tent at the tip of the twig.) In due course, an adult moth emerges from the chrysalis and the cycle begins again.

The Spruce Budworm is the most destructive of all conifer insects, causing enormous losses in pulpwood forests. Though not the only species

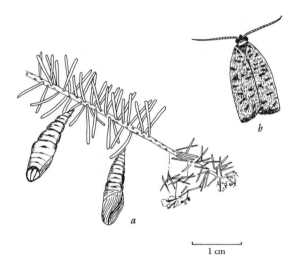

1 cm

Figure 67. Spruce Budworm. (*a*) Two chrysalids (pupae); (*b*) adult moth.

of budworm, it is the most abundant. Some closely related species attack Jack Pines, and others the western spruces.

Another leaf-eating caterpillar that does serious damage is the Larch Casebearer (*Coleophora laricella*). It is the larva of a tiny (wingspan, 1 cm) grayish moth. Indeed, nearly all pest caterpillars grow to be inconspicuous, undistinguished small moths; one of the few exceptions is described later.

The caterpillar of the Larch Casebearer[2] is also small (half a centimeter when full grown), but even so, an infested larch tree can be recognized instantly. Many of its leaves are colorless at the tip (see Figure 68*b*), and some of them have additional colorless leaf tips sticking out from them at right angles (Figure 68*c*). Examination with a lens, which needn't be strong, reveals what is going on. The casebearer caterpillar "mines" the larch leaf. It cuts a neat, exactly circular hole in the leaf (as in Figure 68*b*) and eats all the soft, green tissue inside as far as it can reach. Only the outermost layer of cells—the epidermis of the leaf—is left undamaged, except at the entry hole. Incidentally, the colorlessness of this layer demonstrates to the observer without a microscope that the leaf epidermis of larches lacks chlorophyll, as is true of all the conifers.

Once it has hollowed a leaf tip, the casebearer lines it with silk and then cuts it off for use as a "case"; it also cuts off the extreme tip of the case so that its tubular home is open at both ends. It then lives in, or half in, the case for the rest of its immature life, until it forms a pupa. It wanders slowly over the leaves, dragging the case with it, and whenever it finds a succulent

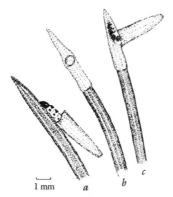

Figure 68. Larch Casebearer.   1 mm   *a*   *b*   *c*

one it cuts a circular hole and begins to feed. While it feeds, the front of the caterpillar is inside the leaf being mined and its rear is protected by the case. Only on its journeys from one leaf to another is any part of the caterpillar exposed, as in Figure 68*a*.

Like so many pests in North America, the Larch Casebearer is not native. It was introduced from Europe at the end of the 19th century. Like many introduced insects, it flourished in its new home because the parasites and disease organisms that attacked it in its original home had been left behind; consequently, its numbers have increased so much that it has become a significant pest. A larch can withstand defoliation better than an evergreen conifer because it is adapted to grow a new crop of leaves every year, and if it is defoliated in spring, it can grow a second crop of leaves in a single season. Even so, of course, several defoliations in succession will kill any larch tree.

A caterpillar with a far less flimsy protection against enemies and weather is the Pitch-nodule Maker, otherwise known as the Northern Pitch Twig Moth (*Petrova albicapitana*). It attacks Jack Pines and Lodgepole Pines. The adult of the species, a small, nondescript moth, lays its eggs at the base of the needle sheaths of the pines, and the caterpillars that hatch from the eggs bore into the twigs and eat the soft inner bark.

Between the time it hatches from the egg and the time it forms a pupa, the caterpillar constructs three blisters, or nodules, one after another, and of increasing size, to accommodate it as it grows. Figure 69 shows the third and largest of the nodules made by its occupant. The earlier, smaller nodules are flatter and more blisterlike.[3]

To decide whether an apparent blister does indeed contain a caterpillar, you need only cut it open. If it is a true pitch blister, the caterpillar will be found inside, completely imbedded in pitch or resin. Even when the trees

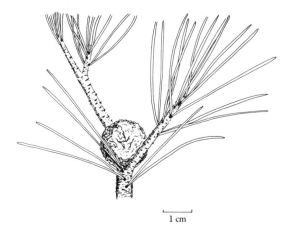

1 cm

Figure 69. A pitch nodule.

attacked by this pest are not killed by it, they are seriously injured; young trees become deformed.

The final lepidopteran pest to be considered here is the Pine Butterfly (*Neophasia menapia*). It is a western species whose caterpillars eat the leaves of Western White, Ponderosa, and Lodgepole pines and also those of firs and Douglas-fir. They are seldom abundant enough to do much damage, but when they are, they can strip a tree completely of its foliage and are therefore an ever-present threat.

The caterpillars are green and rather ordinary. It is the adults that make the species noticeable. Unlike the small, dingy moths into which most conifer caterpillars develop, they are conspicuous butterflies. They are not strong fliers, and flutter feebly high among the trees, where their whiteness makes them conspicuous. Occasionally, an individual will be found sucking nectar from flowers at ground level. The pattern is black on white, as

1 cm

Figure 70. Pine Butterfly.

shown in Figure 70; the color contrast is slightly less distinct in females, which also have dark markings on the front edge of the hind wing (concealed, in the drawing, by the rear edge of the fore wing). Moreover, in females the black pattern on the lower side of the hind wings is outlined in brilliant orange.

SAWFLIES

In addition to "ordinary" leaf-eating caterpillars as conifer pests, and very like them to look at, are the leaf-eating pseudocaterpillars. These two entirely different groups of insect larvae can be told apart by the number of pairs of *prolegs* they have. Prolegs (or abdominal legs) are the paired fleshy bumps on the lower side of a caterpillar or pseudocaterpillar, behind the three pairs of true legs. All caterpillars—pseudo as well as real—have three pairs of true legs, but pseudocaterpillars have a larger number of prolegs. They have six or seven pairs, whereas there are at most five pairs on a genuine caterpillar (see Figure 71).

If that were the only difference it would hardly deserve comment. But pseudocaterpillars are, in fact, entirely different from ordinary caterpillars. Whereas ordinary caterpillars become moths or butterflies when they reach maturity, pseudocaterpillars become sawflies, insects belonging to the same zoological order, the Hymenoptera, as bees, wasps, and ants.

There are many kinds of sawfly.[4] Those that damage conifers belong to two genera. The several species of pine sawflies are all rather alike and belong to the genus *Neodiprion*; they attack pines of all kinds and, as well, one of the species is a serious pest of Western Hemlock. The Larch Sawfly (*Pristiphora erichsonii*), in a different genus, is not closely related to the pine sawflies. Figure 72 shows an adult female pine sawfly.

A pine sawfly is rather like a hefty wasp, without the tiny waist charac-

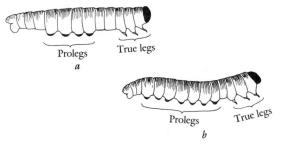

Figure 71. (*a*) Caterpillar (becomes moth or butterfly); (*b*) false caterpillar (becomes sawfly).

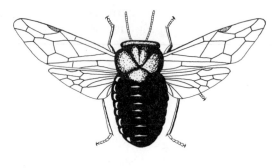

1 mm   *Figure 72.* Pine sawfly.

teristic of a true wasp. Sometimes a swarm of pine sawflies will be encountered buzzing around energetically. As they are fairly large insects, it is a relief to know that they do not sting. The female has an egg-laying organ (*ovipositor*) consisting of two sawtoothed blades (hence the name *sawfly*) that she uses to make a slit in a pine needle so that she can lay her eggs inside. The pseudocaterpillars that hatch are usually gregarious, and masses of them can often be found, shoulder to shoulder, on the leaf tufts of infested pines. The reproductive arrangements of pine sawflies are like those of most other Hymenoptera; the female stores the male's sperm and can lay either fertilized eggs, which produce females, or unfertilized eggs, which produce males. She produces fewer sons than daughters—sons are, after all, less useful to the species than their sisters—and therefore males are less often seen than females. They are smaller than the females and are faster fliers as they are not weighed down with eggs. Males are most easily distinguished from females by the fact that they have feathery antennae; the antennae of females are plainer, as in Figure 72.

Some pine sawfly pseudocaterpillars are active in spring before the current year's crop of new leaves has expanded; they eat the old leaves, stripping the branches bare, and when the new leaves do expand, they form isolated tufts at the tips of the leafless branches. In other species the pseudocaterpillars hatch from the eggs later in the season; they feed in the summer, starting with the newest leaves and then moving on to the older ones. When summer ends, pseudocaterpillars of all kinds form tough silken cocoons, which lie buried in the carpet of fallen pine needles beneath the trees all through the winter. They are safe from the cold, but provide a feast for shrews, which eat great numbers of them.

The Larch Sawfly differs from the pine sawflies in several ways. It is not so big, and has a broad orange band around the abdomen. Males of the species are very rare indeed; they make up only about 2 percent of the population. They are useless, as well as rare. In this species, reproduction proceeds by *parthenogenesis*, that is, by virgin birth with no mating.

BEETLES

The beetles (Coleoptera) form the largest group of insects in terms of number of species; it is believed that about 40 percent of all insects are beetles. Not surprisingly then, many of them depend on coniferous trees for their food. Of those that catch the attention of naturalists, the most noteworthy are the bark beetles[5] (several species of the genus *Dendroctonus*) and the engraver beetles (several species of the genera *Ips* and *Scolytus*). The immature beetles, which are grubs, feed on the soft tissues of the innermost bark of many coniferous trees, leaving behind the tortuous grooves known as *galleries* that are found when the bark of a dying tree flakes off; the galleries show up both on the outer surface of the wood and the inner surface of the bark. Some species construct galleries with beautifully regular herringbone patterns; others have galleries that wander around with no particular pattern. One of each kind are shown in Figure 73 (but note that some *Dendroctonus* species make herringbone galleries like those of *Scolytus* and *Ips*. The *Dendroctonus* gallery shown was made on Lodgepole Pine by the Mountain Pine Beetle, *Dendroctonus ponderosae*).

Experts can tell from the shape of a gallery the species of the beetle that made it. The galleries are started by the adult beetles, which bore into a tree from the outside. The females then lay their eggs; the little notches in the *Dendroctonus* beetle's gallery in Figure 73*b* show where individual eggs were laid. When the eggs hatch, the grubs extend the tunnels, feeding and growing larger as they advance; therefore the tunnels become wider with increasing distance from the site of the egg.

These beetles are divided into so-called primary and secondary attackers. Primary attackers are beetles that attack healthy, undamaged trees, where-

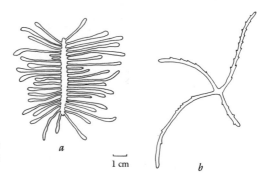

Figure 73. Galleries. (*a*) Engraver beetle (*Scolytus*); (*b*) bark beetle (*Dendroctonus*).

a

1 cm

b

as secondary attackers attack only trees that have already been weakened by primary attackers. The *Dendroctonus* beetles are primary attackers, and *Ips* and *Scolytus* beetles are for the most part secondary attackers. Thus *Dendroctonus* is a far more serious pest, and its many species do tremendous damage. In the East are species that attack Tamarack and White Spruce, but it is in the West, where various species attack all the pines, all the spruces, Douglas-fir, and Western Hemlock, that bark beetles are of particular concern. They are as important a pest in the West as Spruce Budworm is in the East.

Huge areas of Lodgepole Pine forest are often devastated. A forest ravaged by *Dendroctonus* beetles is a mess. The naturalist has no difficulty recognizing the cause of the devastation—there are too many unmistakable signs to leave any doubt—but it is interesting to see how many of the signs can be observed on the dead and dying trees.

First, of course, are the galleries, which are visible where bark has fallen off; often there are other, much wider galleries as well, made by larger beetles that lay their eggs on trees already dead or dying. Chief among these are the Flatheaded Borers, grubs which as adults are the lustrous and often brilliantly colored Metallic Beetles. As Flatheaded Borers' galleries are much wider and more winding than those of *Dendroctonus* beetles, the two kinds are easy to tell apart.

Trees attacked by *Dendroctonus* are sometimes scarred and the scars are distinctive.[6] Unlike scars caused by fires, their lower ends are often some distance above ground level, and they may also be discontinuous (as in Figure 74*a*). Bulging ropelike ridges of scar tissue frequently develop (see

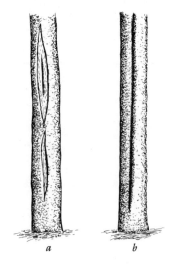

5 cm

a   *b*

Figure 74. Marks left by *Dendroctonus* beetles. (*a*) Open scars; (*b*) ropelike calluses.

Figure 74*b*). Patches of dead bark are often peppered with small black "shot holes"; these are the exit holes left by emerging adult beetles. Once a tree is almost or quite dead, its bark falls away in large, irregularly shaped pieces, leaving bare, white patches visible from a distance. Although they could be called scars, these patches are entirely different from the long, narrow, smooth-edged scars described above.

When a tree is attacked by comparatively few beetles, so that it is not quickly killed, it secretes quantities of resin which fills the beetles' entry holes, making so-called *pitch tubes*. These appear as numerous pale yellow dots and patches on the outside of the bark, often with resin dribbling down from them. The name pitch in this context means pale yellow resin, not pitch-black tar.

Among the most interesting of the diagnostic signs of *Dendroctonus* attack are the blue stains sometimes found on the wood where sheets of bark have flaked off. The color comes from the microscopically small cells of the Blue Stain Fungus, living in the wood. It is believed that the beetles and the fungus are dependent on each other. The fungus is spread by the beetles, which carries its spores from tree to tree. The beetles benefit because infection with the fungus kills some of the tree's tissues and reduces the quantity of resin that the tree can produce, improving the beetles' chances of survival and successful reproduction, as a heavy flow of resin will drown them. (It is interesting to note parenthetically that Dutch Elm Disease is caused by a fungus spread by bark beetles).

A heavy flow of resin is the tree's only defense against *Dendroctonus* beetles and it works only when the attackers are few. If there are only a few attacking beetles, the tree does not immediately succumb but remains healthy enough to produce sufficient resin to drown the attackers. An attack that is unsuccessful because the attackers are overwhelmed by resin is called by foresters a *pitch-out*. To overcome a tree's defenses, the beetles therefore attack en masse; by doing so, they manage to kill their tree victim so quickly that it has no time to produce resin in useful amounts. The problem for the beetles is to communicate so that they can combine their efforts. The signals they use are *pheromones*, chemicals whose scent attracts hordes of other beetles (of the same species) from considerable distances. The beetle army can then become powerful enough to overwhelm the defenses of its target tree.

Except when they are marshaling for an attack, the adult beetles are seldom seen, because, like the grubs, they live inside their galleries. The adult beetle is surprisingly small for a species that causes so much havoc. The largest species of the genus is 7 millimeters (0.3 in.) long at most. All the species look very much alike. That in Figure 75 is *Dendroctonus pon-*

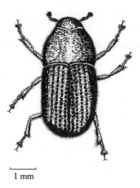

1 mm

Figure 75. Bark beetle (*Dendroctonus*).

derosae, the Mountain Pine Beetle. *Dendroctonus* beetles are more damaging to living conifers than any other kinds of beetles. Although other beetles pale into insignificance beside them as pests, there are many other interesting beetles that catch the eye of a naturalist. The Metallic Beetles (Flatheaded Borers), with burnished wing cases that look as though they were made of metal, have already been mentioned; their larvae are flathead grubs that bore into dead or dying trees.

A number of weevils (or snout beetles) are also pests. Weevils as a group are easy to recognize; a weevil is a beetle that has its mouth at the tip of a long, downwardly curved snout and its antennae attached to the sides of the snout. An important conifer pest weevil is the White Pine Weevil (*Pissodes strobi*), which damages young Eastern White Pines, and occasionally Jack Pines and spruce. It is an eastern species, like the trees it feeds on; out West, a closely related weevil damages young Lodgepole Pines.

White Pine Weevils seldom kill the trees they attack; they usually kill only the leader (the terminal shoot of the stem, above all side branches) with the result that one or more side branches grow large and take the leader's place. If only one side branch takes over, the tree will have a crooked trunk, with a kink where the side branch replaces the original leader; if two or more side branches become leaders, the result is a forked tree, which is unmerchantable.

The way the weevils operate is as follows: The female lays her eggs in a hole in the bark at the tip of the leader, and when the grubs hatch, they tunnel down under the bark, eating living tissue as they go. What makes the process remarkable is that the members of a family of grubs often act in unison; they range themselves side by side heads downward so that they form a complete hollow cylinder just inside the bark, and in that formation tunnel their way downward side by side. Not surprisingly, the leader is killed.

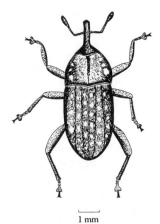

Figure 76. White Pine Weevil.   ⊢—⊣
1 mm

The last beetles to be described here are the Sawyer Beetles; there are several species, all in the genus *Monochamus*. They don't attack living trees, but their grubs feed on the wood of freshly cut logs, leaving boreholes as large in diameter as a pencil (which, of course, reduce the value of the wood). What brings the grubs to the attention of naturalists is the sound of their feeding. You are most likely to hear it if you go into recently cutover conifer forest while the logs are still on the ground. If the logs have been there for any length of time, they will almost certainly contain immature Sawyer Beetles, voracious grubs with no legs but strong jaws. The steady, rhythmical sawing sound of their chewing is easily heard from as far as 10 meters. On a hot, windless summer afternoon, when the birds are silent and (except to a naturalist) the cutover land seems lifeless, the only evidence of active life may be the sawing sound of the Sawyer grubs, steadily chewing wood with the relentless regularity of metronomes.

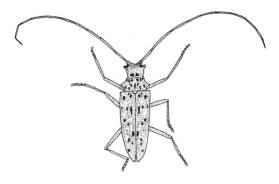

Figure 77. Sawyer Beetle. (Fifty times as big as the weevil above.)   ⊢—⊣
5 cm

Adult sawyers are large beetles and have exceedingly long antennae; for this reason the family they belong to are known collectively as the Long-horned Beetles. The males have antennae twice as long as those of the females, making it easy to distinguish the sexes. Figure 77 shows a male Northeastern Sawyer (*Monochamus notatus*).

BARK LICE

The last group of insects to be considered here are the bark lice, tiny (1 mm long), soft-bodied, sap-sucking insects. They are similar in many ways to aphids or plant-lice (the green fly so familiar to gardeners) and used to be classified as aphids. Nowadays they are treated as forming a separate insect family, known as adelgids. Unlike aphids, which bear living young, the adelgids reproduce by laying eggs.

The insects themselves are too small to attract attention; it is the deformities they cause that make them noticeable. If spruces are found with the tips of many of the twigs tightly curled (see Figure 78), it is because the trees are infested with adelgids which have formed curved galls. More about these galls later.

Another sign of infestation by adelgids is often seen on firs, especially in the East. A Balsam Fir attacked by the Balsam Woolly Adelgid (it is more familiar by its old name, the Balsam Woolly Aphid) is covered with patches of white "wool," actually masses of tangled wax threads that protect the soft insect from dehydration and from enemies.

0.5 m

Figure 78. Spruce tree with spruce galls.

The spruce galls and the fir "wool" are all formed by various species of the bark louse genus *Adelges*. Two species account for the spruce galls: galls on western spruces are the work of *Adelges cooleyi*, the Cooley Spruce Gall Adelgid; and those on eastern spruces are made by *Adelges abieti*, the Eastern Spruce Gall Adelgid. The wool on firs is the work of *Adelges piceae*, the Balsam Woolly Adelgid; it is a very serious pest of Balsam Fir in the East and in the West attacks Amabilis Fir. (Note that the specific name *piceae* applies to pests of the tree genus *Abies*, the firs; and the specific name *abietis* applies to pests of the tree genus *Picea*, the spruces. Confusion existed in the past and cannot now be easily put right. A name change is more troublesome than a misleading name in scientific work.)

The biology of the adelgids is exceedingly complicated. Many species infest two different species of host tree alternately; they live for one or more generations on a spruce, and then move for another series of generations to a different host, which may be a fir, pine, hemlock, larch, or (in the West) a Douglas-fir. Galls always form on an infested spruce, but not on the alternate host tree, which the adelgids injure in some other way. The manner of reproduction is not the same in every generation either. Most generations are produced parthenogenetically; the whole population consists of females, who lay eggs that hatch into more females; males play no part in the process. Then, a two-sex generation will appear, and ordinary sexual reproduction will take place and lead to a new all-female generation, which then reproduces itself parthenogenetically for another few generations. The whole process is then repeated.

The Balsam Woolly Adelgid is a pest that immigrated into North America from Europe early in the 20th century. It is a "degenerate" adelgid, in that its life cycle is less complicated than that of its evolutionary ancestors; it sticks to one host tree, which is always a fir, and it never has a sexual generation. All members of the species are female and they reproduce parthenogenetically for generation after generation, apparently without end. A young adelgid, newly hatched from an egg laid on a fir branch, crawls around until she finds soft tissues where she can insert her sharp *stylet* (feeding tube) and feed. Once she has succeeded and has started sucking up the tree's juices she never moves again, but remains in one spot for the rest of her life, feeding, growing, laying eggs, and finally dying; her life is as sedentary as that of a barnacle. While feeding, she injects into the tree a toxic chemical, which causes swellings to form. We are reminded of the effect of a mosquito bite, but in the case of a fir tree the swelling (*gouting* to foresters) is permanent. Severely infested trees also change color; they become a rusty red and soon die.

Now for the gall-forming adelgids. The two common ones are the

1 cm   *Figure 79.* Spruce galls.

Eastern Spruce Gall Adelgid and the Cooley Spruce Gall Adelgid, as mentioned above. The first of these is not an important pest; although infested eastern spruces may be covered with galls because of it, they seem to suffer no ill effects. The Cooley adelgid, however, is a serious pest and does much damage to the western spruces (on which it forms galls) and also to its alternate host tree, Douglas-fir (on which it doesn't form galls). The mechanics of gall formation are as follows: A batch of tiny newly hatched adelgids move to the end of a growing twig and each one selects a young leaf for itself and begins to suck sap at the base of it, causing the leaf base to swell and to form a protective roof over the insect. Because all the leaves of a twig are attacked, the separate small swellings coalesce into a single large swelling, a gall. When the adelgids grow to egg-laying age, they emerge separately from the gall, which is left full of holes, as in Figure 79. When the dead, dwarfed leaves have all fallen from a gall, it looks like a deformed cone rather than a deformed twig and galls are often mistaken for cones. From a distance, a gall-covered spruce appears to have its twigs curled at the tip, as shown in Figure 78.

OTHER INSECTS

We could add indefinitely to the list of conifer pest insects. Many of them are interesting and some of them have fascinating names. There are the Lodgepole Needleminer, the Blue Horntail, the Saratoga Spittle Bug, the Nantucket Pine Tip Moth, the Agile Pine Needle Aphid, the Hemlock Looper, the Pine Tortoise Scale, and the Bagworm (a caterpillar, needless to say).

Most of the insects we see in conifer forest are up to no good or are, at

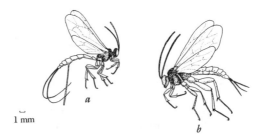

Figure 80. Two ichneumons.   1 mm

best, harmless. There are plenty of beneficial insects too, but most of them are small and seldom seen. An insect is described as beneficial, it need hardly be said, when it benefits mankind, for example by destroying pest insects.

Insect pests can be killed by other insects in two ways. Sometimes the beneficial insect eats the pest. For example, the Dubious Checkered Beetle (and some related species) eat bark beetles. In other cases the beneficial insect parasitizes the pest. There are numerous species of small wasps that do this. In most cases the parasitic wasp finds, stings, and lays an egg in the larva of the pest insect. The wasp egg hatches and the grub grows and develops within the pest larva, eating it from the inside as it develops. By the time the pest larva has been reduced to a dried husk, the wasp grub is ready to form a pupa, from which, in due course, an adult wasp emerges. Then the cycle repeats.

Most parasitic wasps are small and inconspicuous. Those most likely to be noticed are the ichneumons,[7] which are obviously related to hornets and wasps, though they are smaller and have thinner abdomens. Two are shown in Figure 80. On the left is *Scambus tecumseh*, a parasite of the Spruce Budworm, and on the right is *Theronia atalantae*, a parasite of the Pine Butterfly.

Forest entomologists make great efforts to foster beneficial parasites, and to find and introduce new ones from overseas. If the destruction of a pest insect can be left to its insect enemies—a process known as biological control—it means that the pest can be eliminated without the use of insecticides. It all sounds so easy. In practice, of course, there are difficulties; only occasionally does a parasite species become so abundant that it reduces the population size of the pest it parasitizes to a level at which the damage done by the pest is negligible. This is scarcely surprising. If the parasite did too thorough a job in eliminating its host, it would itself die out.

NOTES

1. S. A. Graham and F. B. Knight, *Principles of Forest Entomology* (New York: McGraw-Hill Book Company, 1965).

2. Department of Forestry and Rural Development, *Important Forest Insects and Diseases of Mutual Concern to Canada, the United States and Mexico* (Ottawa, 1967, out of print).

3. A. H. Rose and O. H. Lindquist, *Insects of Eastern Pines* (Ottawa: Canadian Forestry Service Publication No. 1313, 1973).

4. A. H. Rose and O. H. Lindquist, *Insects of Eastern Pines* (Ottawa: Canadian Forestry Service Publication No. 1313, 1973).

5. J. B. Milton and K. B. Sturgeon (eds.), *Bark Beetles in North American Conifers* (Austin: University of Texas Press, 1982).

6. R. G. Mitchell, R. E. Martin, and J. Stuart, "Catfaces on Lodgepole Pine—Fire Scars or Strip Kill by Mountain Pine Beetle?" *in Journal of Forestry*, vol. 81, pp. 598–601, 1983.

7. L. A. Swan and C. S. Papp, *The Common Insects of North America* (New York: Harper & Row, 1972).

Chapter 7

Fungi and Other Parasites

No tree lives forever. The commonest causes of death among the conifers, if we exclude deaths due to logging, are insect pests (described in Chapter 6), fire (to be described in Chapter 8), and disease, the chief topic of this chapter. It may seem morbid to harp on disease and death; what is a disaster for a tree, however, brings life to countless insects and disease-causing fungi. In any case, disease and death are impossible to overlook; diseased and dead trees will be found in any conifer forest—except, perhaps, in young plantations—and the naturalist who ignores them is missing a topic of tremendous interest. Birders and other naturalists who specialize in animals of one kind or another seldom give much thought to the subject. The reason is that diseased animals do not long survive and their remains are quickly disposed of; carnivores weed out the weak and ailing, and scavengers ranging from bears, coyotes, and crows to maggots and molds soon clear away dead bodies. Among plants, too, soft-tissued herbaceous plants soon disappear after they have died and withered. That leaves trees. In a natural conifer forest, trees affected by rusts, blights, diebacks, rots, galls, cankers, stains, and witch's brooms are to be found on all sides.[1] Who has not seen the ailments shown in Figure 81? On the left is a canker, caused by a parasitic fungus; on the right is a burl, caused by an unknown agent, possibly bacteria.

Before describing some of the more easily recognizable infectious diseases of conifers, some common noninfectious afflictions deserve mention. It is not generally realized that trees, like people, are susceptible to sunburn and frostbite. If the trunk of a tree that has been well shaded is suddenly

Figure 81. (*a*) Canker; (*b*) burl.

exposed to the hot summer sun, which can happen if adjacent trees are felled, the bark may be injured. It reddens, and may turn a bright red, on the southern, sunlit side; and if the burn is severe, the reddened bark becomes scaly and flakes off; the parallel with human sunburn is close.

Frostbite happens if a sudden early frost in the fall freezes fresh bark tissue before a tree has had time to harden off; the parts that are frozen and killed are eventually sloughed off, and scar tissue grows over the wound. A frostbite scar is difficult to recognize as such; it looks much like any other scar. Although an early frost in the fall can cause frostbite on a tree's trunk, a late frost in spring is damaging too, especially to trees growing in frost pockets—hollows in the land into which cool air drains on calm spring nights. Frost kills the succulent new shoots which tend to curl, and the frost-killed leaves turn reddish brown. Usually it is hard to tell whether the injury was caused by frost or by a fungus disease.

Now for the infectious diseases.

Mushrooms and Bracket Fungi

There are plenty of mushrooms and bracket fungi to be found in conifer forests, especially in the fall. The bracket fungi are those that grow out like shelves or ledges from the sides of tree trunks.

As remarked in Chapter 5, a mushroom is the above-ground, spore-producing part of a fungus whose permanent "body" lives underground in the form of a network of delicate filaments called *hyphae*; the network itself is known as the *mycelium* of the fungus. Thus a mushroom is to a fungus what a flower is to a flowering plant. Many of the mushrooms to be found in conifer forests, perhaps the majority, grow up from fungi whose hyphae unite in the soil with the roots of the trees to form mycorrhizae (see Chapter 5). These fungi therefore benefit the trees; indeed they are essential to the trees' well-being. Very few of the mushroom-producing fungi cause disease. A group of species that does belongs to the genus *Armillaria*. This genus contains several closely related species collectively known as Honey Mushroom; the name refers to the color, not the taste, though these mushrooms are edible. They are also called Shoestring Fungus, for a reason that will become clear below.

The Honey Mushroom does not always cause injury. It can live harmlessly on the roots and stumps of dead trees; and it sometimes infects a living tree without doing any damage. The trees most likely to suffer from a Honey Mushroom "attack" are those that are already sickly from some other cause, and they may be severely damaged. Most conifer species are susceptible. The fungus causes the bark and wood of the base of the trunk and the roots to rot (hence yet another of its names, Root Rot Fungus).

When a tree is infected, its growth slows, its leaves turn yellow and fall, and it may produce a "distress crop" of cones; in anthropomorphic terms, conifers (like many flowering plants) often make a desperate effort to leave abundant offspring when death looms. These signs of ill health are not peculiar to Honey Mushroom attacks, of course; they merely show that a tree is sick. The diagnostic signs are these: Quantities of resin flow out of an affected tree from the base of the trunk, and there may be so much that it soaks into the dead leaves and soil surrounding the tree and congeals there to form a crust. If the loosened bark at the base of the tree is pulled off, you find a layer of white "felt," actually a mycelium of very densely tangled hyphae; the whole layer is usually shaped like a fan. If the infection is of long standing, rotted bark at the bottom of the trunk may have fallen away, leaving a scar, and the exposed fan of mycelium will have dried up and become black. Lastly, you find the "shoestrings" (technically, *rhizomorphs*). These are thin, dark brown cords consisting of numerous parallel strands of hyphae enclosed in a dark outer layer. A shoestring grows

5 cm

Figure 82. Honey Mushroom (Shoestring Fungus).

through the soil from an infected tree until it encounters the root of a healthy neighboring tree; it is able to penetrate a healthy root, and once it has done so, new hyphae sprout from its tip and grow through the tissues of the new tree victim. This is one of the ways the fungus spreads.

The fungus also spreads by spores, formed on the gills of the mushrooms that grow in clumps around the bottom of an infected tree (see Figure 82). Mushrooms do not necessarily appear around an infected tree every year; the fungus may fail to produce any mushrooms for several years in a row, but when they are formed they usually appear in the fall.

The great majority of the fungi that cause a conifer to decay grow entirely inside the tree and do not send rhizomorphs through the soil to neighboring trees as the Honey Mushroom does. They are the bracket fungi. The mycelium of such a fungus stays inside the tree, with its hyphae twining in and among the tree's bark and wood cells and drawing nourishment from them. When the time comes to grow a spore-producing body, the fungus grows what amounts to half a mushroom cap, with no stalk, that protrudes from the tree's trunk like a shelf or a wall bracket. These shelves or brackets are known to foresters as *conks*, and when they are ripe they produce clouds of spores. The spores find their way into other trees via unhealed wounds such as woodpecker holes or the stubs of wind-snapped branches; in this way, decay spreads from tree to tree.

There are numerous species of bracket fungi, and each of them has a distinctive conk and causes an equally distinctive type of decay or rot. Thus there are red heart rot, brown stringy rot, red ring rot, brown cubical trunk rot, white butt rot, white mottled rot, brown cubical pocket rot, brown trunk rot, long-pitted rot, white spongy trunk rot, brown crumbly rot, and many, many more. Their conks range from tiny, thin, leaf-shaped structures to hard, leathery "shelves." The conk of the fungus that causes brown cubical trunk rot is the Sulfur Fungus (*Laetiporus sulphureus*); its fresh conks are a brilliant orange-yellow, and are fleshy and edible making them a favorite with fungus collectors. The fungus causing brown cubical butt rot (as distinct from the corresponding trunk rot) is Velvet Top Fungus (*Phaeolus schweinitzii*), so called because of the bright chestnut-red "velvet" on its upper surface; it is also known as Dye Polypore, because a good dye can be obtained from the conks. They usually grow at ground level, or not far above.

Two other frequently encountered conks are shown in Figure 83. The conk of Quinine Fungus (*Fomitopsis officinalis*) is hard and hoof-shaped. It derives its popular name from its extremely bitter, quininelike taste; indeed, attempts were made at one time to culture the fungus artificially, so that a pharmaceutically useful chemical could be extracted. The conks are

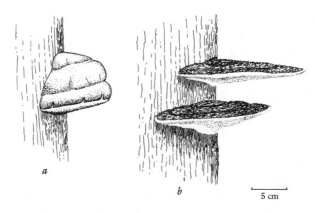

Figure 83. (*a*) Quninine Fungus; (*b*) Artist's Fungus.

especially likely to be found on larches, but the fungus attacks other conifers as well. The mycelium part of it causes brown trunk rot, which is very destructive.

The conk of Artist's Fungus (*Ganoderma applanatum*) is one of the commonest of the shelf fungi (it is sometimes called, simply, "shelf fungus"). The reason for its popular name is that if you scratch its white undersurface with a pointed stick, the scratch appears immediately as a sharp, dark brown line; on a large conk, it is possible to write a long message, or draw a detailed map. It attacks many tree species, hardwoods as well as conifers, and its mycelium causes white mottled rot.

As remarked above, there are a tremendous number of different species of wood-destroying fungi. Many of them attack only trees that are already dead, and cause them to decay. Others attack the dead heartwood of living trees or even living sapwood. Some confine their attacks to only a few species of victims, while others attack a wide variety. Some can grow only on wood that has already begun to decay from other causes, whereas others can infect a living tree and then continue to flourish on the same tree after it is dead and to consume the wood until decay is complete.

The damage done by these fungi in exploitable forests is tremendous. They cause enormous losses to the lumber industry. But in a wilderness ecosystem their presence is indispensable to the well-being of the forest community as a whole. By bringing about the disintegration of old, dead trees, they restore nutrients to the soil and enable new trees to be nourished by their predecessors. If it were not for the decay fungi, nutrients would remain locked up in standing or fallen dead trees while seedlings starved; growth would slowly come to a stop until a fire wiped the slate clean and the cycle of growth could begin again. In our cool, northern area, fire is usually more important than decay in keeping the cycle going and preventing stagnation; even so, the fungi play a very significant part.

THE RUSTS

A large group of pathogenic (disease-causing) fungi are known as the rust fungi or, more simply, just as the "rusts." They are all parasites that must have living plants as hosts, unlike the decay fungi described in the preceding section, most of which can live on dead trees. Their spore-producing organs, instead of being mushrooms or shelf fungi, are patches of pustules or blisters growing on the surface of the host plant; sometimes, if they happen to be reddish brown, they look vaguely like patches of rust. With a very few exceptions, they all parasitize two unrelated plants (or unrelated groups of plants) alternately. Thus they resemble such insects as the Spruce Gall Adelgid, which alternates between spruces and Douglas-fir (see Chapter 6). Finally, again in contrast to the decay fungi, they tend to be fairly host-specific; that is, one species of rust can parasitize only a few closely related species of conifers as one of its hosts, and only a few closely related species of plants (a different set, of herbs, shrubs, or hardwood trees) as its alternate host.

Many of the rusts cause serious losses in our forests. The most serious is White Pine Blister Rust,[2] which will attack all four of the five-needle (soft) pines in our area, that is, the two white pines (Eastern and Western) and the two stone pines of the mountain regions, Whitebark and Limber pines. When a pine forest is swept by an epidemic, the results are devastating. Most of the young trees infected with the fungus are killed. Therefore, although older trees may manage to survive, few of their offspring grow to maturity, and in time, when the old trees have died off, nearly all the pines in the forest are gone. Soft pines are the coniferous hosts of this rust—its Latin name is *Cronartium ribicola*—and its alternate host is shrubs of the genus *Ribes*, the currants and gooseberries, or, collectively, ribes. There are numerous species of ribes and nearly all are suitable as alternate hosts, the cultivated as well as the wild.

When the trunk or branch of a pine tree is infected, it swells and becomes discolored. Eventually, as the fungus grows and consumes its host's living tissues, the tree becomes girdled. A young tree is almost certain to die; in older trees, if the infection is at some height up the trunk, only the top of the tree, above the level at which it has been girdled, dies; the lower part survives. Trees with dead tops, as in Figure 84, are known as *staghorns* and are a common sight in rust-infected forests. After several

Figure 84. Western White Pine (top killed by White Pine Blister Rust).   ⌐ 1 m

years of life in a pine, the rust fungus is ready to produce spores. In the spring, numerous white blisters break out on the bark of the infected tree and in these the spores are formed. Ripe spores are a bright orange, and the white outer covering of the blisters splits open to release them. The breaking blisters are interesting and colorful when examined under a lens, but this is of little comfort to an observer who knows that a great many of the soft pines in a region may well be doomed. The tiny spores float

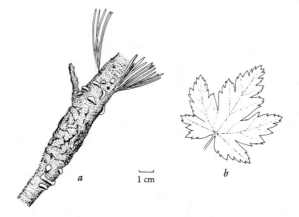

Figure 85. (*a*) White Pine Blister Rust; (*b*) ribes leaf.

through the air for considerable distances and those that land on ribes leaves grow and prosper. The rust does hardly any damage to the ribes. It quickly grows to maturity and produces spores that can infect only other ribes. Through the summer, generation after generation of the rust on ribes produce more ribes-infecting spores; it takes only a couple of weeks or so for a spore landing on a ribes plant to mature and produce new spores of its own, and in this way the disease spreads quickly among all the ribes plants in an area. Then in the fall the rusts on the ribes plants produce spores capable of infecting pine trees again, via the pines' leaves.

White Pine Blister Rust is by far the worst of the rusts, but there are scores of others. Many other species of the genus *Cronartium* cause rusts. They have other conifers as one of their hosts, and other shrubs or herbs as alternate hosts. All the *Cronartium* species cause "blister rusts"; some are named for the appearance of the blisters, others for the alternate host. Several blister rusts attack the hard pines. One, Stalactiform Blister Rust, has Indian Paintbrush (various species of *Castilleja*) as its alternate host in the West, and Cow-wheat and Yellow-rattle as alternate hosts in the East. Another, Comandra Blister Rust, has Bastard Toadflax, a species of *Comandra*, as alternate host. Still another of the blister rusts that attacks the hard pines is Sweet Fern Blister Rust, whose alternate hosts are Sweet Fern and Sweet Gale. It would be easy to multiply the examples of blister rusts, but there are numerous other rusts to mention, among them leaf rusts, cone rusts, and rusts that cause witch's brooms and galls.

Take witch's brooms first. These are the densely bushy masses of proliferating branches often seen on a conifer. They are obviously pathological. They have a variety of causes, and infection by a rust fungus is only

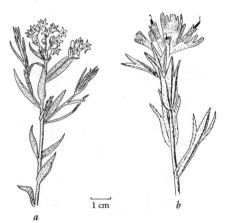

1 cm

a   *b*

Figure 86. (*a*) Bastard Toadflax;
(*b*) Indian Paintbrush.

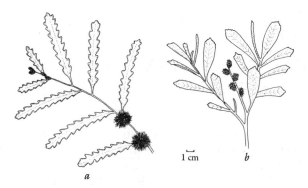

Figure 87. (*a*) Sweet Fern; (*b*) Sweet Gale.

one of the possibilities. Two genera of rust fungi cause their host trees to grow golden yellow witch's brooms, which contrast strikingly with the dark-green foliage of the healthy parts of the tree. These brooms also contrast with witch's brooms due to other causes (to be discussed later), which usually have leaves of the same color as the rest of the tree. One of the rusts causing a color change is the Spruce Broom Rust, *Chrysomyxa arctostaphyli*, which causes dense yellow brooms to grow on spruces (see Figure 88). Its alternate host is the creeping shrub Kinnikinnick, otherwise known as Bearberry (*Arctostaphylos uva-ursi*), on which it produces purplish brown leaf spots. Conspicuous yellowing of spruce leaves is also occasionally caused by *Chrysomyxa ledicola*, for which the alternate host is Labrador Tea, of the genus *Ledum*.

The other common rust that causes yellow witch's brooms, this time on fir trees, is *Melampsorella caryophyllacearum*; various species of chickweed serve as alternate hosts.

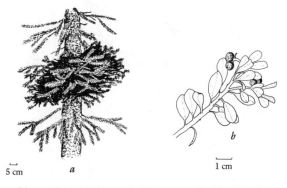

Figure 88. (*a*) Yellow witch's broom; (*b*) Kinnikinnick.

Returning to *Chrysomyxa*: Besides the species that cause yellow witch's brooms on spruces there are also some species of the genus that affect spruce cones; their alternate hosts are the wintergreens, species of *Pyrola* and *Moneses*. Cones infected with rust produce no seeds; they are discolored and somewhat deformed and, at times, are covered with a dust of spores.

The leaf rusts exist in bewildering variety. They produce the white pimples like miniature fingers that are often found on the leaves of needle-leaved conifers. Figure 89*b* shows them on the leaf of a fir. This one has willow as its alternate host. Figure 89*a* shows the effect of rust infection on a scale-leaved conifer, in this case Eastern Juniper ("eastern red cedar" to the lumber trade). The growth, which appears in wet weather in spring, is called a cedar apple; it has numerous gelatinous "fingers." It is caused by one of the many species of the genus *Gymnosporangium*, and has apple and crabapple trees as alternate hosts. In some fruit-growing regions in the eastern part of our area, the rust is an important disease of apple orchards. Thus there is an interesting contrast between Cedar Apple Rust and White Pine Blister Rust. In White Pine Blister Rust it is the coniferous host, the pines, that are commercially valuable; the alternate hosts, at least when they are wild ribes plants, are of no concern. In the case of Cedar Apple Rust, it is the hardwood host, orchard apple trees, that are the commercially valuable victims; by comparison the scattered Eastern Juniper trees that are the coniferous host are of no great value.

Another species of *Gymnosporangium* causes witch's brooms on Rocky Mountain Junipers, and its alternate host is Saskatoon Berry.

The final rust to consider here is one of the very few that does not alternate between two unrelated hosts. It confines itself to the hard pines and is very destructive. In addition to killing trees, it stunts and maims

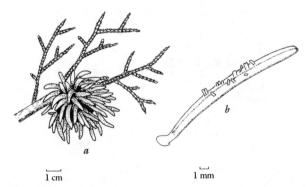

Figure 89. (*a*) A cedar apple on juniper; (*b*) rust on a fir leaf.

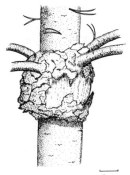

Figure 90. Gall caused by Western
Gall Rust.

5 cm

many others. It is known as Western Gall Rust,[3] although it occurs all
across the continent, and is caused by *Peridermium harknessii*. The eye-
catching symptom of the disease is the crop of galls on the trunk and
branches of an infected pine; there may be scores of them on a single tree.
There are often witch's brooms as well, usually just above the galls. The
bark on each big, spherical gall flakes off to expose the smooth wood
underneath, except where collars of dead bark remain, at the top and
bottom of the gall.

We could continue ad infinitum. There are literally hundreds of different species of rusts. Each has one or more coniferous host species, and
(with a very few exceptions) one or more alternate host species. Each
causes a distinctive set of symptoms on both kinds of hosts: swellings,
discolorations, blisters, pustules, witch's brooms, cankers, galls, and peculiar growths such as cedar apples. They would provide a happy hunting
ground for amateur naturalists if microscopes (optical and electronic) were
as easy to use as binoculars. Perhaps that day is not far off, if the evolution
of home computers is any guide. In the meanwhile, the rusts are easy to
find and interesting to look at. Even if you cannot name them and can do
no more than guess at their alternate hosts, it is worth the effort to become
familiar with the conifer rusts in your own stamping grounds.

SOME MOUNTAIN SPECIALS

Some tree diseases of the western mountains deserve special mention,
because any naturalist who lives among, or visits, the mountains will
almost certainly come across them sooner or later.

The commonest is Douglas-fir Needle Blight, which is due to a fungus
(*Rhabdocline pseudotsugae*) that attacks only Douglas-fir. It causes the
Douglas-fir leaves to become mottled and spotty; the spots are yellow in

1 cm

Figure 91. Needle blight of Douglas-fir.

the winter and become brownish red as the season advances. Finally, the blighted leaves drop off. A severely affected tree that after several seasons loses all its leaves will die, of course; milder attacks merely have the effect of slowing a tree's growth. Young trees are particularly susceptible, and therefore the disease is easily observed by a naturalist; the mottled leaves are low down and visible, not out of reach in the canopies of full-grown, mature Douglas-firs. The disease also attacks plantation-grown Douglas-firs in the eastern part of our area.

A less common disease, of unknown cause, is Cork-bark of Alpine Fir. A short section of the fir's trunk, near the bottom, develops very thick ridges of cork with deep furrows between (see Figure 92*a*). The disease doesn't appear to be lethal. An Alpine Fir afflicted with cork-bark is not to be confused with so-called Corkbark Fir, which is a geographic variety of the common Alpine Fir that grows only in the Southwest, in New Mexico, Arizona, and southern Colorado, far outside our area. Corkbark Fir always

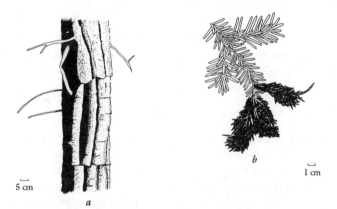

5 cm

b

1 cm

a

Figure 92. Two diseases of Alpine Fir. (*a*) Cork-bark; (*b*) Brown Felt Blight.

has soft, corky bark; its thick cork is natural, not a sign of disease. Another disease of Alpine Fir, and also of Engelmann Spruce and (occasionally) Rocky Mountain Juniper, is Brown Felt Blight (Figure 92b). There is another Brown Felt Blight that attacks pines, especially Whitebark and Limber Pines high in the mountains. Both diseases are caused by fungi, two different fungus species on the two sets of hosts. The two felts look much the same; the fungus grows all over the leaves at the ends of the twigs and glues them together in a stiff, dirty mass. These felts are unattractive but interesting. They grow best under the snow, and are therefore particularly likely to be found on low, ground-hugging branches. The reason that they are confined to high altitudes is probably that up there the snow, which protects the fungus, lasts all winter; at low altitudes fungal felts would probably not survive when a thaw melted the insulating, moisture-retaining blanket of snow.

DWARF MISTLETOE

Fungi are not the only parasites on conifers. There is also a group of parasitic flowering plants that attack them, the dwarf mistletoes.[4] To most people a mistletoe is a bushy, leafy, green plant, with white berries, which grows on the branches of deciduous trees and which provides the branches hung in doorways at Christmas to mark a compulsory kissing station. These are not the only mistletoes, however. There are also dwarf mistletoes that parasitize only conifers. They are related to ordinary mistletoes (same family, different genus) but are quite unlike them to look at.

The dwarf mistletoes, as their name implies, are exceedingly small plants, easily overlooked unless deliberately sought for. The place to start searching is on a tree with one or more witch's brooms, as dwarf mistletoe is much the most frequent cause of brooms. The leaves on the brooms match those on the rest of the tree, unlike those of the yellow, rust-caused brooms described earlier. A luxuriant green witch's broom, like that in Figure 93, may give the impression that the tree it grows on is flourishing but that is definitely not the case. Trees that bear brooms are severely injured by them. The leaves on a broomed tree (on both the broom itself, and on the rest of the tree) will often be found to be shorter, and paler in color, than those of an unparasitized tree. Therefore, as it has less chlorophyll, the affected tree must obviously be growing more slowly than its unparasitized neighbors.

Trees with witch's brooms die prematurely, and in infected forests dead, leafless trees with dead, leafless witch's brooms on their trunks and branches are common. In many forest areas, especially in the West, dwarf

⌐1 m

Figure 93. Witch's broom on a Lodgepole Pine.

mistletoe is more injurious than all the fungus-caused diseases put together.

The plants that cause all this damage belong to the genus *Arceuthobium*. They can be found growing out of the bark of the trunk or branches of an infected tree in little tufts; the trunk or branch is often (not always) swollen. The dwarf mistletoe plants are yellowish green to brownish green; sometimes they are bushy, as in Figure 94 (drawn in winter), and sometimes their stems grow singly from the trunk of the host tree. There are several species of *Arceuthobium* in our area, one in the East and the

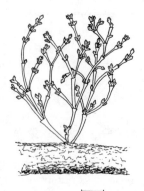

1 cm

Figure 94. Dwarf Mistletoe.

others in the West. They are not identical, of course, but the differences among them are slight, and the genus is instantly recognizable once you have seen any member of it. The eastern species is *Arceuthobium pusillum*, common on the eastern spruces. Among the western species are *A. americanum*, the largest, which grows on Lodgepole Pine (this is the one shown in the drawings); *A. douglasii* which grows on Douglas-fir; and *A. campylopodum* which grows on a large number of hosts: firs, spruces, pines, hemlocks, and junipers.

The visible part of a dwarf mistletoe is only part of the plant; growing down from it, into the tissues of the host tree and parasitizing them, are long strands of tissue that do the job that in a nonparasitic plant is done by the roots. The purpose of the visible tufts growing into the open air is to produce pollen and seeds. These are formed on different plants; that is, a dwarf mistletoe is either a male plant whose flowers produce pollen but lack seeds, or a female plant whose flowers bear seeds but produce no pollen. Figure 95 shows part of a plant of each sex.

The mechanism by which the ripe seeds are discharged from the female plant is a masterpiece of evolution. Each fruit (the oval objects on the plant in Figure 95*b*) contains one seed. As the fruit ripens, the stalk supporting it lengthens and curves over, until it is pointing downward. Then, at the same instant as the fully ripe fruit falls from the stalk, its outer skin contracts sharply; the seed inside is shot upward, in the same way that a piece of wet soap will shoot upward if you suddenly squeeze it. The flying dwarf mistletoe seed can reach a speed of 80 km (50 mi) per hour, and can land more than 10 m (30 ft) from its starting point.

Dwarf mistletoes have some natural enemies, but there are not enough of them to be useful as control agents. There are a few fungi that parasitize dwarf mistletoes, a case of a parasite parasitizing another parasite.

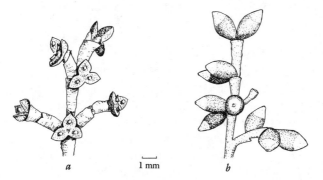

a   1 mm   b

Figure 95. Dwarf Mistletoe. (*a*) Male plant; (*b*) female plant.

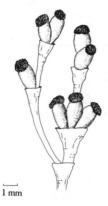

1 mm

Figure 96. Dwarf Mistletoe parasitized by a fungus.

One of these fungi (shown in Figure 96) is *Wallrothiella arceuthobii*. Curiously, the only part of the dwarf mistletoe that it will grow upon is the female flowers; it destroys the seed of each flower it attacks, and its own spore-producing organs look like miniature backberries perched on the dwarf mistletoe fruits.

Some Nonparasites

Three herbaceous plants that are often found growing in the shade of coniferous trees are sometimes mistaken for parasites on the conifers' roots. In fact, the roots of these plants hardly ever make contact with the trees' roots, but confusion is natural because the plants are never found anywhere except in the shade of conifers; and, as they are nongreen and thus obviously devoid of chlorophyll, they must be obtaining their food ready-made from some source outside themselves; they cannot photosynthesize.

The three plants occur throughout our area. The commonest of them is Indian Pipe (*Monotropa uniflora*), so-called because of its shape. As shown in Figure 97*b*, the flower hangs down when it first opens, giving the plant the shape of an old-fashioned clay pipe. The plant is also known as Ghost Plant because, when fresh, its fleshy stems and flowers are pure white; the leaves are mere scales. As the plant ages, the nodding flower turns until it faces upward, and the whole plant turns black. The second plant of the trio is Pine Sap (*Monotropa hypopitys*); it differs from Indian Pipe in color—the whole plant is golden orange—and in having several flowers on each stem rather than just one. The third plant is Pine Drops (*Pterospora andromeda*); it is much taller than the other two, is reddish brown, and is covered with sticky hairs. Many of the dead flowering stems remain upright throughout

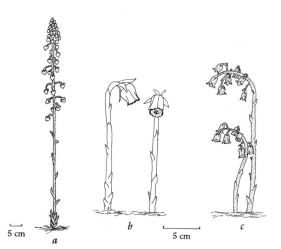

Figure 97. (*a*) Pine Drops: (*b*) Indian Pipe; (*c*) Pine Sap.   5 cm   *a*   *b*   5 cm   *c*

the winter; instead of withering or rotting after they have shed their seeds, they dry and harden where they stand.

The way in which these three plants obtain food is most unusual. The mystery was unraveled by the Swedish botanist Erik Björkman.[5] It turns out that the plants share the mycorrhizal fungi of the conifers near which they grow. As described in Chapter 5, nearly all conifer trees have mycorrhizae; that is, each tree's roots are covered with, and penetrated by, the hyphae of a fungus which obtains its food (sugars and other carbohydrates) from the tree and, in return, serves to extend the tree's root system through a larger volume of soil than the roots themselves can reach. When one of the three herbs begins to grow near a tree with mycorrhizae, hyphae from the mycorrhizal fungi are attracted to the herb's roots by a chemical that the herb exudes. The fungal hyphae grow around, and penetrate, the herb's roots in the same way as they do the tree's roots and form a bridge between tree and herb. The herb seems to obtain at least some of its food from the fungus, which had it in the first place from the tree. The fungus benefits too; it obtains a growth-stimulating substance from the herb.

This complicated arrangement was unraveled by Björkman's experiments. He worked with Pine Sap growing close to spruce trees. The existence of a link between a tree's roots and a herb's roots via the tree's mycorrhizal fungus was first suspected when both sets of roots, and the fungus, were dug out of the soil and seen to be connected. The linkage was not itself proof that carbohydrates pass from tree to herb through the fungus. The evidence that this is indeed what happens was obtained by injecting radioactive carbon into the bark of a spruce and discovering that nearby Pine Sap plants had become radioactive five days later. It does not follow that the herb obtains all its nourishment in this way. It may also, in part, feed as a *saprophyte*, that is, feed on the remains of dead organisms of all kinds in the soil. But in any case, it is not a direct parasite on conifer roots.

NOTES

1. J. S. Boyce, *Forest Pathology* (New York: McGraw-Hill Book Company, 1961).
2. Y. Hiratsuka and J. M. Powell, *Pine Stem Rusts of Canada* (Ottawa: Canadian Forestry Service Forestry Technical Report No. 4, 1970).
3. R. E. Foster and G. W. Wallis, *Common Tree Diseases of British Columbia* (Victoria: British Columbia Forestry Branch Publication No. 1245, 1969, out of print).
4. J. A. Baranyay, *Lodgepole Pine Dwarf Mistletoe in Alberta* (Ottawa: Canadian Forestry Service Publication No. 1286, 1970).
5. E. Björkman, "*Monotropa hypopytis* L., an epiparasite on tree roots," *in Physiologia Plantarum*, vol. 13, pp. 308–27, 1960.

Chapter 8

The Elements: Fire, Wind, Snow, and Air Pollution

Many fates befall the conifers. Some of the insect pests that attack them were described in Chapter 6, and some of the fungi that infect them were described in Chapter 7. Now we come to the forces of inanimate nature that beset them: fire, snow and ice, and wind. By far the most important of these is fire.

FIRE

Fire is as big a danger to a coniferous tree as insects and fungi. A fair comparison is difficult however. The various risks are not independent of one another, because fire is far more likely in a forest containing numbers of dead and dying trees—the victims of bark beetles or spruce budworms or shoestring fungus, perhaps—than in a green and healthy forest. In any case, from the point of view of an individual tree, fire is a serious and ever-present threat.

From the point of view of a whole forest, however, fire is a blessing. If it were not for fires, the remains of dead vegetation would accumulate year after year and the forest floor would become covered with an impenetrable tangle of fallen trees, broken branches, and withered and slowly rotting vegetation of all kinds. Dead trees and their parts contain all the mineral nutrients they obtained from the soil while they were growing, and these nutrients remain locked up inside them, inaccessible to other trees, until the dead remains are decomposed and returned to the soil, either by decay or by fire. In our cool northern latitudes, decay caused by fungi and bacteria proceeds too slowly to keep up with the continually replenished supply of dead material; hence the need for fires, to keep the forests as a whole growing vigorously.

In wet regions, where fires are infrequent, partly rotted material does indeed accumulate. The bulk of it consists of dead mosses and fallen conifer leaves; as layer is added to layer, the bottom layers, which are cold, soggy, and airless, become peat. The process is called *paludification*. Fires break out occasionally, and even burn away some of the peat. But in many parts of our area, especially in northern Minnesota, eastern Manitoba, and northwestern Ontario, there are places where the peat forms faster than it burns or decays. The process has been going on for thousands of years, since the end of the warm period that followed the melting of the ice sheets at the end of the last ice age. These ancient but still growing peat bogs provide a poor soil for trees and support little more than slow-growing black spruce. They owe their distinctive characteristics to the fact that they burn so seldom. In most forests, fires are far more frequent.

Forest fires are of many different kinds. The commonest are surface fires, as in Figure 98, which burn everything flammable lying on the forest floor but which do not seriously injure full-grown trees with thick, corky, fire-resistant bark such as Red Pine in the East and Ponderosa Pine in the West. The open Ponderosa parklands of the West owe their existence to frequent surface fires; the fires consume fallen dead branches and kill young seedling pines, but the older pines are uninjured and have enough space to grow tall and stately, as they are free of competition from lower vegetation. Most of

Figure 98. Surface fire.

the pines are fairly fireproof. Thin-barked trees, such as the firs and spruces, are much more sensitive; a mild fire kills them even when they are fully grown.

The most impressive fires are crown fires. Sometimes a forest fire burns only through the trees' crowns (as in Figure 99). Other fires are a combination of surface and crown fires; everything above ground burns, and if large quantities of fuel (dead trees, branches, snags, and logs) have accumulated since the last fire, the result is a devastating holocaust in which winds generated by the fire itself can become strong enough to snap the trunks of large trees.

A third type of fire is the ground fire. Ground fires take place only where there is an accumulation of peat. All the action is subterranean and the only evidence that a fire is in progress is provided by occasional wisps of smoke emerging from small holes in the ground and the pleasant, if disturbing, smell of smoldering peat. A ground fire started by a campfire that was not completely doused can proceed a long way under the surface without changing the appearance of the ground in any way. When this happens, the normal-looking forest floor may be only a thin roof over large, blackened caverns where the underlying peat has been gradually burned away. Ground fires are also started by surface fires. They can continue unobserved for months, smoldering slowly and invisibly below innocent-looking carpets of pine needles.

Figure 99. Crown fire.

Besides differing from one another in type (surface, crown, and ground), forest fires vary greatly in their intensity and therefore in the ease with which they can be contained (if it is desirable to contain them). The fiercest fires are those fanned by strong winds following weeks of dry weather, in forests containing large amounts of fuel. The quantity of fuel, the accumulation of dead snags, stumps, logs, and branches waiting to burn, is a very important factor. That is why it is not necessarily wise to fight all fires. A mild surface fire that is not a threat to camps, cottages, and towns, and that is clearing away a modest amount of accumulated fuel, is best left to burn itself out. If it is needlessly extinguished, the chance to burn off accumulated trash harmlessly is lost, and the next time fire strikes there may be so much fuel that the fire becomes fierce and dangerously uncontrollable. A fierce fire creates new fuel as well as disposing of old fuel. Trees that a fire kills are usually not consumed but remain as standing dead snags awaiting the next fire.

Fire is a natural event. In wilderness country, fires are started by lightning and have recurred more or less regularly ever since the land became forest-covered. Indeed, there is a so-called fire cycle; any patch of forest land will be burned over sometime, whereupon a new forest will grow up and will be burned in its turn, and so on ad infinitum. The duration of the cycle, the interval from one fire to the next, is determined only partly by chance. The average duration in many forests is of the order of 100 or 200 years; it is much shorter (about 25 years) where the climate is dry and forests of pine (Jack in the East, and Lodgepole in the West) grow on thin, sandy or rocky soils; it is longer (more than 400 years) in the rain forests of the Pacific Northwest, and longer still in muskegs and wet peat bogs in the Far North. These figures are only averages. What will happen at a given locality at a given time depends on local conditions and can only be predicted, with much uncertainty, by a person with considerable local knowledge and experience.[1]

A fire does not necessarily remove all the accumulated fuel at one fell swoop, of course. Large chunks of wood may persist through several fire cycles, only small fractions of their mass burning away in any one fire. This is especially true in the West Coast rain forests, where many trees are huge and downed logs are constantly wet.

For many forests, fire is a necessary part of their life cycles. For example, parklands of Ponderosa Pine cannot remain as they are indefinitely without regular fires, as mentioned above. If fires are prevented, the parklands are replaced in time by dense forests of Douglas-fir. The fact that fires are of crucial importance in maintaining particular kinds of forests has led to the theory (suggested by the forest biologist R. W. Mutch) that some species of trees have evolved not merely to survive fire but actually to promote it. In other words, certain trees, those that thrive only in sunlit, open places, actually require fairly frequent fires to ensure their survival. A forest of such trees is called a fire-dependent forest.[2]

The argument is as follows: A fire-dependent species will clearly benefit if its wood is highly flammable, because high flammability will make fires more likely. As a result, trees of these species have evolved characteristics that ensure they will catch fire more easily, and burn more hotly, than trees of species that are not fire-dependent. The evolution of these characteristics presumably takes place by the process of natural selection. Trees that have the desirable (selectively favorable) characteristics are more likely to leave offspring than those that lack the characteristics, as their seeds have a better chance of landing in open, burned-over land. Therefore if the characteristics are heritable, trees that possess them will become commoner, and trees that lack them will become rarer, until in time every living tree of the species is of the highly flammable, hot-burning kind.

Trees that survive a hot fire are seldom wholly unscathed. Foresters can deduce the fire history of a forest by examining cut stumps for old fire scars; they can determine when each fire occurred by counting the numbers of annual rings that have grown in undamaged wood of the same tree since each scar was formed. This can be done only with trees that do survive, of course, usually trees with thick, corky bark that insulates the living cambium in the trunk and saves them from death. Thin-barked trees such as firs and spruces are rarely found with fire scars as they rarely survive a fire.

A fire scar is sometimes not very different from a scar due to some other cause. Recognizing the many different kinds of scars to be found in a conifer forest is quite an art. Two common types of scars were described in the chapters on insects and diseases. Scars caused by bark beetles are fairly distinctive; as noted in Chapter 6, bark beetle scars usually start above ground level and often occur in vertical rows of two or three, with gaps between. Scars caused by the Honey Mushroom fungus (see Chapter 7) start below ground level; the wood exposed where the bark has fallen away shows signs of rot at the bottom, and may be covered by a black, dried-up layer of old mycelium. Fire scars, too, usually have their lower edges below ground level and because they are typically triangular, are often called "church-door" scars. But we can only be certain that a scar was caused by fire if signs of charred, blackened wood are still to be seen in the scar itself; charred bits of debris around the scarred tree are suggestive but not conclusive. Another common kind of scar that can sometimes (but not always) be recognized is a bear scar, made where a bear has sharpened its claws on a trunk.

A bear scar is entirely above ground level, at a height on the tree where

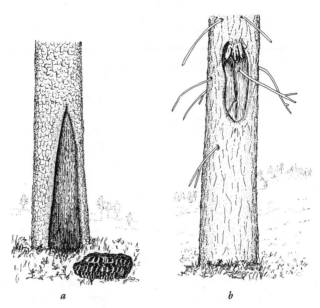

Figure 100. (*a*) Fire scar; (*b*) bear scar.

one would expect a bear to reach to; there may be some tattered strips of torn bark at the top of the scar; and short protruding branches growing out within the scar area (if there happened to be any) are usually still there. Needless to say, the surest sign is a sight of the bear itself; but even without that pleasure, one can enjoy seeing a bear scar and knowing it for what it is. Other scars to look out for are those caused by porcupines (see Chapter 9), and "scrape" scars. The latter are the scars formed when a falling tree scrapes a neighboring tree, usually leaving broken and splintered branches as well as a scraped trunk on the neighbor. The presence of the fallen tree beside the scarred one makes the cause of the scarring obvious.

AFTER A FIRE: SUCCESSION

Wherever a fire has burned, almost as soon as the ash has cooled, new plant life of some sort begins to grow. In time, trees get started and a new forest grows up to replace the burned one. In one sense, every conifer forest can be described as a post-fire forest, as there isn't a hectare of land in our area that hasn't been burned over several times; but the name post-fire forest is usually reserved for the first trees to colonize a burned area. The best-known, most obvious post-fire forests are the dense, monotonous stands of Jack Pine (in the East) or Lodgepole Pine (in the West) that are familiar to every highway traveler (see Figure 101).

These forests are even-aged. All the trees in them began growth at the same time, and because all have grown at roughly the same speed, they are all much the same size—but not of exactly equal size, of course—with the result that the smaller trees are progressively crowded out—deprived of sunlight and water—by the larger. These trees languish and die, and the forest thins itself as it grows. This natural thinning can produce forests so regularly arranged that they look like human handiwork. European visitors have been known to mistake them for plantations. Sometimes, if the trees chance to be very well matched in size, the natural thinning process doesn't work. Instead of a few relatively strong trees succeeding at the expense of the weaklings, enormous numbers of seedling trees keep growing neck and neck. All of them remain weak and stunted because the available sunlight, water, and nutrients are insufficient to support such a throng. When this happens, one finds a "dog-hair stand" of Lodgepole or Jack pines. The phenomenon is particularly common in Lodgepoles. Fifty-year-old dog-hair stands have been found in which there were 25,000 stems per hectare, or about as many "bodies" per unit area as in a closely packed mob of people; the spindly trees were so dense that it was impossible to walk among them.

The reason that Jack Pines and Lodgepole Pines (in the East and the West respectively) are so common on newly burned land is that they are especially well adapted to invade such territory as soon as it becomes available. As remarked in Chapter 3, both of these tree species have serotinous cones, that is, cones whose opening may be delayed for many years. It is the heat of a forest fire that finally causes the large supply

Figure 101. Even-aged Lodgepole Pines.

(perhaps several years' worth) of tightly closed cones on a tree to open and shed their seeds; many are shed from cones at the tops of killed but still standing trees and they land on a layer of cool ashes which forms an ideal seedbed for them. Where the fire cycle is short, as it is in the comparatively dry regions where Jack and Lodgepole pines are so abundant, one pine stand may follow another for generation after generation. Then forests of young seedling pines grow up among the blackened snags of their fire-killed parents, which formed the preceding forests (see Figure 102).

Not all newly burned land is colonized by one of these two closely related pine species. Sometimes the first wave of colonists are hardwoods (see Chapter 10). And there is another conifer, Black Spruce, that retains its seeds in unopened cones for a few years (not for as many years as the two pines) and is therefore in a good position to occupy burned land the moment it becomes available. There are other species that depend on fire to prepare a suitable seedbed for their seeds, particularly Eastern and Western White Pine and Red Pine.

What grows up after a fire depends, of course, on what seeds are available, and this in turn depends on the time of year when the fire happened and on what species of potential parents for the next crop of trees chanced to be within range. Seeds from the two white pines, and also from Red Pine and White Spruce are shed in late summer or early fall, and these species can take possession of new territory only if their seeds fall on newly burned land that hasn't had time to become overgrown with vig-

Figure 102. Young Lodgepole Pines in an old burn.

orous shrubs and herbs that would crowd out conifer seedlings. Therefore a summer fire is a necessary preliminary to their establishment. Furthermore, for one of these species to take over an area, there must be at least one mature parent tree near at hand to provide the necessary seeds; and the potential parents, or at least one of them, must have a good crop of cones in that particular year. No conifer tree bears a heavy cone crop every year; Good cone years come at intervals. For instance, in Red Pines there is often an interval of 5 years and sometimes as much as 10 between one good cone year and the next; Eastern White Pines and White Spruces usually bear a good cone crop every 3, 4, or 5 years.[3] Between times, these trees bear very few cones or even none at all because, unlike Jack and Lodgepole pines, their cones are not serotinous (persistent).

Therefore we often see groups of White Spruces, all of the same size and age, in which a few of the trees have their tops heavily laden with cones (showing that there is nothing wrong with the season) and all the remaining trees, though healthy, are coneless (see Figure 103). The effect is more obvious in White Spruces than in pines because pine cones take 2 years to reach maturity.

Even-aged pine forests are the commonest post-fire forests everywhere in our area except on the West Coast. The coast forests are unique in many ways. Because of the mild climate and high rainfall, the trees—predominantly Douglas-fir, Giant Arborvitae, Sitka Spruce, and Western Hemlock—grow to enormous size and live to great age. Everything is on a grand scale, including forest fires, which tend to be large in area and intensely destructive. The fire cycle here is much longer than in drier climates, the average interval between the large fires being probably be-

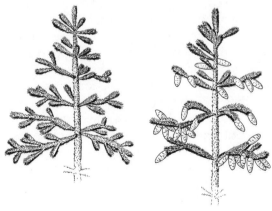

Figure 103. White Spruce tops (same age, same size, same season).

tween 400 and 500 years. The first colonists on burned land are usually Douglas-firs, and they might be expected to form even-aged forests just as the pines do in drier climates, but oddly enough this does not happen. Old Douglas-fir forests are usually far from even-aged; trees of a wide range of different ages are present and only the oldest of them date back to the time of the original fire. The probable cause for this variability is that very small "spot" fires, that are started by lightning from time to time but fail to spread, make "holes" in the forest here and there and the holes provide growing space for new Douglas-fir seedlings. If this happens repeatedly, over several centuries, a mixed-age forest is the inevitable result.[4]

The customary invaders of burned-over land—the pines, Douglas-firs, and also the larches—are known as *intolerant* trees. This means intolerant of shade; foresters use the word by itself, and one is assumed to know without being told what it is that the trees are intolerant of. Conversely, the hemlocks, firs, arborvitaes, and spruces are tolerant. Unlike the intolerant species, which must have sunlight, their seedlings grow well on the floor of a full-grown forest, in the deep shade beneath the canopy formed by the mature trees' crowns. But the tolerant species don't demand shade in the way that the intolerant species demand sunlight; although an intolerant tree cannot thrive in the shade, a tolerant one can grow as well in sunlight as in shade.

Although the notions of tolerance and intolerance in trees are generally thought of as having to do only with light and shade, there is probably more to it than that. It is likely that tolerant species can withstand competition from other trees better than intolerant species can. Tolerant species tend to be less windfirm than intolerant species, however, because their root systems tend to be shallower and less wide-spreading. Thus if surrounding trees are felled, a tolerant tree is more likely than an intolerant one to be blown down. Tolerant trees are normally less exposed to the risk of blowdown, of course, because the surrounding forest shelters them from the wind; moreover they gain additional support from having their roots intertwined with their neighbors' roots.

Intolerant trees cannot invade an existing forest; they can grow only in full sunlight and therefore they must have open land—usually open because of a fire—to get started. That is why the first forest to grow up on a newly burned tract of land is usually composed of intolerant pines or, in the West, of Douglas-fir. What happens next depends on whether the intolerant forest is burned down in its prime (or before) or lives out its full lifespan until the trees start to die of old age. If it burns, trees of the same species can succeed their burned parents: There is a new, burned-over, sunlit area in which they can get started. But if it does *not* burn, offspring cannot succeed parents, since seedlings of an intolerant species cannot grow in the shade of full-grown trees. Instead, young trees of tolerant species, which will already have started to grow in the shade cast by the first, intolerant forest, will grow up to replace the intolerant trees as they die off.

The result is a completely changed forest. A post-fire forest of intolerant pines may be replaced by one of tolerant spruces, firs, and hemlocks. In the West Coast rain forests, a post-fire forest of Douglas-fir may be replaced by one of Sitka Spruce, Giant Arborvitae, and Western Hemlock. These are examples of the process described by ecologists as succession. One kind of vegetation colonizes bare ground, but cannot maintain itself for more than a generation because it alters the environment in such a way that its offspring cannot prosper. The altered environment, however, *is* suitable for another kind of vegetation, which, in the course of time, takes over the ground. One kind of plant community succeeds another in predictable fashion. Thus if we know the requirements of the various tree species to be found in a region, we can foretell, quite precisely, what the course of events will be—that is, what different kinds of vegetation will succeed one another—after a fire has left a tract of ground temporarily bare.

The way in which one generation of trees leaves a suitable environment for its successors is easily seen in the West Coast rain forests. There, the tolerant trees, those that come late in the succession, grow best on a seedbed of rotting wood, the remains of their predecessors. In old forests, where there have been no fires for many hundreds of years, nearly all the conifer seedlings (of tolerant species, of course) grow on rotting logs, or on old stumps if any cutting has been done (see Figure 104), and hardly any on the forest floor. The dead wood on which the seedlings grow are called *nurse logs*.

Figure 104. Western Hemlock seedling on a Douglas-fir stump.

Snow

Except in the mountains, and there only on steep slopes, snow is no menace to conifers. Because of their shapes, the trees usually shed heavy snow loads; the snow slides off before it builds up to branch-breaking weights. Heavy rime ice is another matter, as described in the next section. In the mountains, too, snow is harmless as long as it doesn't slide. The two stone pines (Whitebark and Limber) are renowned for the flexibility of their branches, which bend without breaking when snow weighs them down.

It is sliding snow—avalanches—that do the damage. All mountain travelers are familiar with an avalanche path—a long cleared swath down a mountainside, with dense forest to left and right. The slide scar is often overgrown with young deciduous trees and shrubs, such as aspens, birches, willows, alders, honeysuckles, and many more. Conifer seedlings are absent except perhaps at the bottom where the slope levels off and the avalanches have slowed and lost their power. The deciduous trees and shrubs can recover if their tops are snapped off every few years; provided their roots are undamaged, new shoots can grow up from the broken stumps. As a result, avalanche paths quickly regain their light-green covering of low, deciduous shrubs and tree seedlings. Conifers, on the other hand, cannot sprout from a stump; if a conifer trunk is snapped off, the tree dies.

Figure 105. Avalanche paths.

Figure 106. Pistol-butted tree.

Snow can also have a visible, but quite harmless, effect on the growth of conifers in hilly country where the slopes are too gentle for avalanches. If a heavy, wet snowpack such as often forms in later winter presses against the bottoms of young tree trunks, the trunks acquire a bend that they will keep for the rest of their lives. Because of their curved trunks, such trees are sometimes described as being pistol-butted.

A heavy snowpack is a positive benefit to some conifer species. It weighs down the lowermost branches until they acquire a permanent droop and rest on the ground. Roots develop at each point of contact, and a twig

Figure 107. Layering. (*a*) Winter; (*b*) summer (many years later).

growing up from the prostrate branch directly above these new roots develops into the main stem of a new small tree. In this way an original single tree can develop a whole clone of new trees around it. The process is called *layering*; it is a naturally occurring version of the layering done by nurserymen to propagate such garden shrubs as forsythia. The group of trees produced by natural layering is a clone; that is, all the member trees are genetically identical with one another. Conifer clones often form very dense circular groups of trees with the original founding tree at the center and successively younger trees toward the outside of the circle. Cloning in this way is especially common in Black Spruce and Alpine Fir.

WIND AND ICE

Everyone is familiar with the effect continual strong winds can have on the shapes of conifers. Bent, dwarfed, ground-hugging trees, shaped by a lifetime of fierce winds with few calm days, are known by their German name *krummholz* (crooked timber). An alternative name (a technical term, believe it or not) is *elfin wood*. Krummholz are common along windswept shores and near the tree line on high mountains. Winds laden with sharp particles of ice and snow can sometimes have the effect of making a tree dumbbell-shaped (see Figure 109). The lowermost branches are protected from damage by the snowpack, and high branches are above the abrasive layer of wind-blown snow and ice crystals. Between these immune levels is a zone in which the trunk is repeatedly "sandblasted" (strictly, ice-blasted) until the stumps of broken branches are worn away. The result is a dumbbell-shaped tree.

Trees with this shape are also caused, in sheltered places, by the brows-

20 cm

Figure 108. Krummholz.

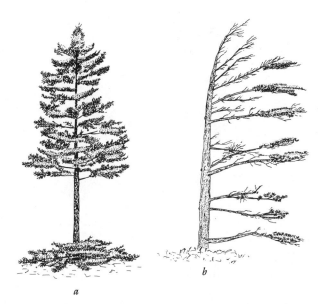

Figure 109. (*a*) Dumbbell tree; (*b*) flag tree.

ing of animals such as deer and snowshoe hare (see Chapter 9). It is usually possible to tell whether ice-blasting or browsing was the cause in a particular case by considering what is most likely; only trees that are exposed to the elements can be ice-blasted. A symmetrical "dumbbell" forms only in places where winds come from different directions in different storms without any one direction being favored most of the time. Where the wind is nearly always from the same direction, that is, where there is a marked prevailing wind, so-called flag trees are formed.

One of the most remarkable effects of wind and ice on conifers was described by the forest ecologist Douglas Sprugel.[5,6] The effect is on the appearance of whole forests rather than on individual trees. In North America the phenomenon has so far been found only in the mountains of New Hampshire and Maine and the affected forests are always Balsam Fir forests. The phenomenon is seen at its best on Mount Katahdin.

What is seen, from the window of a plane or by looking across a valley at the mountain slope opposite, is a series of whitish stripes forming parallel lines in the dark-green forest at intervals of about 100 m (300 ft) or a bit more. The stripes are belts of dead and dying trees. Close inspection, together with determination of the trees' ages, shows that there is in fact an interesting pattern of trees of different ages and conditions, as can be seen in Figure 110*b*, which illustrates how the forest would appear from the

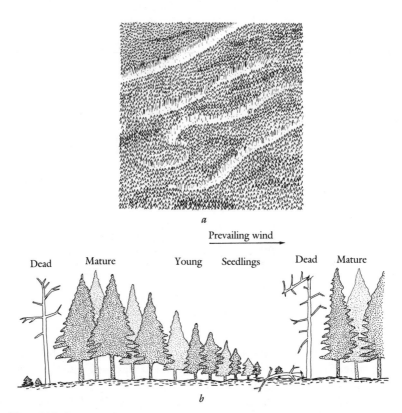

Figure 110. Regeneration waves. (*a*) From the air; (*b*) from the side (diagrammatic).

side if a slice were taken by cutting through it at right angles to the stripes. Note the sequence of trees as you go to the left from the right-hand standing dead tree. The dead tree itself is at the downwind edge of a stripe; immediately to the left of it are fallen dead trees with tiny fir seedlings growing among them. Continuing leftward (upwind), you find progressively older trees, ending with fully mature, living trees, after which the cycle starts again with the standing dead trees that look white from a distance. Observations of these stripes over a period of time showed that the stripes are moving, like the crest of a breaking wave but at the exceedingly slow speed of about 1½ m (4 or 5 ft) per year, in the direction of the prevailing wind (left to right in the diagram). Sprugel's explanation of what is going on in these regeneration waves, as he calls them, is as follows.

Balsam Firs are short-lived trees, and by the time they begin to show signs of old age at around 70 years they are very susceptible to damage and consequent death. The commonest cause of damage, in the northern

Adirondacks, comes in winter, and consists of rime ice plus strong wind. Rime is the ice that forms when a mist of tiny supercooled water droplets (below freezing in temperature, but still liquid) are blown against, and freeze onto, the branches of trees. Rime forms much heavier ice coatings than hoarfrost, which is what appears when invisible water vapor crystallizes out of clear air onto cold surfaces. In a forest with "stripes," the heaviest coatings of rime are formed on the most exposed trees, which are those just to leeward of a belt of low seedlings and saplings. Because they are unsheltered, these trees are also the ones that sway most markedly in a wind, causing their ice-laden branches to snap off. The damaged trees die and in the strip of ground thus left empty, seedling firs begin life when spring comes. In the following winter, the newly unprotected strip of mature trees is exposed to freezing mist and dies. Another strip of seedlings then begins growth to replace them, while the preceding year's strip of seedlings, now a year older and correspondingly taller, is "left behind," by about 1½ m, by the slowly advancing wave. Thus the wave advances downwind, as year succeeds year. These regeneration waves are always found on slopes and they can move either upslope or downslope; which they do depends on the direction of the prevailing wind.

It was remarked above that regeneration waves seem to be found only in Balsam Fir forests. Interestingly, this fact is both a cause and an effect. Waves are much more likely to form in a pure, one-species forest than in a mixed forest of two or more species, because the mechanics of wave formation require that all the trees become decrepit at the same age, and different species differ greatly in longevity. Balsam fir has a life-span of only 80 years or so; by contrast, Red Spruce, which often grows with fir to form mixed forests in these mountains, lives for as long as 300 years. This is why forests of pure fir are necessary for the development of waves. At the same time, the waves themselves cause a fir forest to remain pure; the waves make invasion by spruce (which would make the forest a mixed forest) unlikely.

The way it works is this. In a forest without waves, where there is a mixture of fir seedlings and spruce seedlings, the firs preponderate at first both because firs produce more seeds than spruces do and hence more seedlings, and also because the fir seedlings' roots grow faster and crowd out many of the spruce seedlings. The spruces get their turn to flourish after the short-lived firs die out. Thus in a mixed spruce-fir forest, the firs dominate in the first century but the spruces survive them and dominate once the forest has grown older. In a forest with regeneration waves, no part of it ever grows old. The passage of successive waves over the ground keeps the forest permanently rejuvenated, with conditions that favor fir over spruce constantly recurring. It is no wonder that spruce is soon altogether excluded.

AIR POLLUTION

All naturalists, nowadays, are alive to the menace of air pollution and acid rain. Toxic industrial gases such as sulfur dioxide, which form an invisible ingredient of the air, are more likely than acid rain (rain with the same toxic gases dissolved in it) to harm coniferous trees directly, because the gases enter the interior spaces of the leaves through the stomata along with the air (including some carbon dioxide) that the leaves must have. Trees damaged by air pollution seldom catch the eye of the naturalist, for two reasons. First, the visible injuries caused by severe pollution are not easy to distinguish from injuries caused by the many other stresses that can affect a tree, such as insect attack, fungus disease, or a shortage of water or nutrients. Second, it takes high concentrations of a toxic gas to produce visible damage; low concentrations cause invisible effects, chiefly a slowing of photosynthesis.

Many pollution specialists believe that the best warning of potential pollution damage to a forest is given by the lichens growing on the trees. Because they are small and rootless, and absorb air through every part of their surfaces when damp, lichens are thoroughly exposed to whatever pollutants the air may contain and will quickly show ill effects if they are at all susceptible. Observing the lichens in a forest is therefore one way in which amateur naturalists can attempt to keep tabs on pollution. It is not easy, however. Learning to recognize the many common lichen species takes much time and patience. Moreover, although the presence of lichens on the trees in a forest is an encouraging sign so far as air purity is concerned, the opposite is not true. A forest in perfectly clean air is often quite devoid of lichens and the trees probably benefit from their absence. It is the disappearance of lichens from places where they had been abundant, rather than their absence in places where they have never been found, that arouses suspicions of pollution.

Three of the species used for pollution monitoring are shown in Figure 111. All three are common everywhere in our area and are often to be found on the trunks and branches of conifers. *Hypogymnia physodes* is pale gray and puffy, and is loosely attached to the tree it grows on; *Parmelia sulcata* is whitish gray; *Evernia mesomorpha* is bushy, with rather flabby greenish yellow "branches."

The effect of acid rain on conifers is hard to judge. The direct effect is probably negligible; what gives rise to concern is the indirect effect, caused

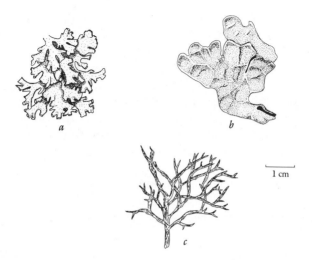

Figure 111. Three common lichens that grow on conifers. (*a*) *Hypogymnia physodes;* (*b*) *Parmelia sulcata;* (*c*) *Evernia mesomorpha.*

by the action of acid rain on the soils in which coniferous forests grow. The damage that acid rain may be doing to the soil is a matter of much debate. Obviously, soils differ enormously in their sensitivity; a thin, acid soil is likely to show ill effects much sooner than a deep, alkaline soil, which may be able to absorb and neutralize large quantities of acid rain. The crucial question, concerning any given soil, is this: What is the maximum rate at which the soil can absorb and neutralize acid without itself being changed? The answer sought is not a quantity but a *rate.* Soil is not inert; it contains vast numbers of living microorganisms, and multitudes of chemical reactions are constantly in progress, at least in unfrozen soil. Therefore a soil can dispose of contaminants by itself, without allowing them to accumulate, provided they do not enter the soil too fast. The question becomes: How fast is too fast?

Consider an analogy. Suppose, in a dark kitchen, you turn on a faucet over a sink that has an open drain hole, of unknown size, at the bottom. As long as the rate of flow from the faucet is slow enough, the water will flow out through the drain as fast as it flows in from the faucet; it will not fill the sink. But if the flow from the faucet is speeded up to the point at which it exceeds the rate at which water can flow out through the drain, the sink will gradually fill and overflow. In the dark, you will be unaware that the sink is filling until the overflow begins.

The parallel with acid rain is obvious. The size of the drain hole (the capacity of different soils to neutralize acid rain) is not known. As the kitchen is dark, we cannot tell how fast the sink is filling; it may overflow, unexpectedly, at any time. Likewise, our forests may at any time be devastated because imperceptible damage has been accumulating. Ecological research, by its very nature, yields results slowly, far too slowly to warn of impending danger in time for it to be headed off. As an example we have only to look at the very severe damage that acid rain has unquestionably done to freshwater ecosystems in the eastern part of our area; these ecosystems succumb more quickly than forests do to destruction by acid rain. Research did not prevent this damage, although it has, most convincingly, explained it.

A similar scenario is only too likely in the forests; irreparable damage may suddenly appear, while research plods along. We must expect the worst and adopt the only wise course—the cleaning up of air pollution.

NOTES

1. D. C. West, H. H. Shugart, Jr., and D. B. Botkin (eds.), *Forest Succession: Concepts and Applications* (New York: Springer-Verlag, 1981).

2. R. W. Mutch, "Wildland Fires and Ecosystems—a Hypothesis," *in Ecology*, vol. 51, pp. 1046–51, 1970.

3. T. T. Kozlowski and C. E. Ahlgren (eds.), *Fire and Ecosystems* (New York: Academic Press, 1974).

4. D. C. West, H. H. Shugart, Jr., and D. B. Botkin (eds.), *Forest Succession: Concepts and Applications* (New York, Springer-Verlag, 1981).

5. D. G. Sprugel, "Dynamic Structure of Wave-regenerated *Abies balsamea* Forests in the North-eastern United States," *in Journal of Ecology*, vol. 64, pp. 889–911, 1976.

6. D. G. Sprugel and F. H. Bormann, "Natural Disturbance and the Steady State in High-altitude Balsam Fir Forests," *in Science*, vol. 211, pp. 390–93, 1981.

Chapter 9

Warm Blood:
The Mammals and Birds
of Coniferous Forests

Coniferous forests provide a distinctive habitat for mammals and birds. There is deep shade all the year round, and shelter from blizzards in winter. There is often an abundant supply of seeds from the trees themselves for seed-eaters, an equally abundant supply of insects for insect-eaters, and a rather more chancy supply of prey for the carnivores that eat both the seed-eaters and the insect-eaters. There is usually a shortage of ground vegetation, except in clearings, because of the dense shade. Even so, hiding places for ground-dwelling animals are not hard to find. Concealment is also easy for the tree dwellers as the treetops are leafy all through the year.

Not surprisingly, any naturalist who searches for them can find numerous birds and mammals in conifer forests. Some of these are wide-ranging species that spend part of their time in different habitats, but there are others that are found only among conifers. It is this latter group, the warm-blooded animals for whom conifer forests are a lifelong home, that provide most of the subjects for this chapter. But before we come to them, we should mention two important pest mammals that damage conifers, and deciduous trees as well.

Two Mammal Pests

Far the most serious of all the pests that damage coniferous forests, as every naturalist well knows, are members of our own species. The destruction of forests in our area by human agency is devastating. The sight of clear-cut land, once forested but now covered with mounds of slash, is depressingly familiar; as is the sight of landslides and eroded gullies, and rivers made turbid by the mud-laden waters that flow down deforested slopes. Naturalists from another planet would probably be fascinated to watch the way in which some human beings deliberately destroy their own habitat while other human beings, who share the habitat, look on with dismay. The gravity of the situation varies from region to region within our area, and depends on the intelligence or lack of it of local governments. The subject has been carefully and thoroughly chronicled in numerous recent books and articles.[1,2]

Turning to a more attractive animal species that is at the same time a pest in a small way, consider the porcupine, a beast that endears itself to everybody except dog-owners and foresters. It is a pest only in winter, when it damages trees by eating the bark; it seldom does this in summer, when fresh leaves and other succulent greens are available. Where porcupines are numerous the damage they do to conifers is considerable. When fresh, the damage is conspicuous and easily recognizable. There are large, irregularly shaped patches of gnawed wood on the trunks and branches of the trees. The grooves left by the porcupine's teeth make it obvious that the barkless patches are the work of an animal; and if the patches are so high up that they must be the work of a tree-climber, as opposed to a ground-dweller like a moose or an elk, then we can be sure that they are porcupine scars. Even gnawed scars at a low level are more likely to have been made by porcupines than by anything else if they are on conifers.

When members of the deer family gnaw bark, they prefer deciduous trees such as aspen. Patches without bark high on a pine tree can also be a sign of bark beetle attack (see Chapter 6), but evidence of bark beetle injury, where whole sheets of bark have flaked off, is found only on pines that are obviously moribund, and the exposed wood lacks all signs of tooth marks.

The aftereffects of porcupine feeding may be a ruined tree, at least from the forester's point of view. Porcupines often girdle a tree high up; the upper part of the tree, above the level of the girdling, then dies and the result is a so-called staghorn tree. Porcupine damage is not the only cause of staghorning, however; any injury that kills the top of a tree and leaves the lower part healthy creates a staghorn, for example, the girdling of pines by White Pine Blister Rust (see Chapter 7). Another habit of porcupines is to eat the terminal buds of seedling trees, which kills the leader and causes side branches to grow vertically upward and to thicken, thus creating a forked or many-trunked tree (see Figure 112). Anything that kills the terminal bud of a young tree has the same effect, of course. A forked tree is not a sure sign of porcupine damage, though it certainly arouses suspicion.

Figure 112. (*a*) Porcupine gnawing bark; (*b*) forked tree injured as a seedling by a porcupine.

OTHER MAMMALS: BROWSERS AND SEED-EATERS

In winter, life is difficult everywhere in our area, in coniferous forests as well as in other habitats. The problem is food shortage: the lack of grasses and fresh, succulent leaves. Large vegetarian animals have to make do with the harsh, dry foliage of conifers, and with twigs, buds, and bark from trees and shrubs of any kind; they have to browse when they cannot graze. The small vegetarians are seed eaters.

Most members of the deer family (Mule and White-tailed Deer, elk, and moose) obtain their winter food by browsing. Caribou are an exception; they survive on a diet of lichens.

The browsers browse on both the evergreen conifers and on deciduous trees and shrubs depending on what is available, except that elk are said not to like spruce. The two species most dependent on evergreens are Mule Deer in the West and White-tailed Deer right across our area. Mule Deer in the West Coast rain forests seem to be particularly fond of browsing on the seedlings and saplings of Douglas-fir and Giant Arborvitae. In the dry interior country of the West their staple is Ponderosa Pine. White-tails, too, browse on a variety of conifers, but when they can, they concentrate on Eastern Arborvitae.

The other important browser is the Snowshoe Hare. Though Snowshoe Hares are at large in the forest all through the year, naturalists see them far more often in winter than in summer, even though their coats turn white in winter to match the snow. Their big snowshoelike hind feet leave conspicuous tracks in the snow, and even though the hares are less active

Figure 113. Mule Deer.

during the daytime than at night, they can sometimes be seen hopping about (not very strenuously) in patches of undergrowth. Their white coats then show up against the dark background of the undergrowth; or if the background is snow, their eyes or the black tips of their ears make them noticeable. In winter they browse on all the twigs they can reach, and when hares are numerous, the lower branches of trees and shrubs may show a clear "browse line." The height of the browse line depends, of course, on the height of the snow surface on which the hares stand. As the snow deepens, successively higher, not-yet-nibbled branches are brought within reach and by the end of winter the browse line may be high up. Deer, of

Figure 114. Snowshoe Hare.

course, do not have this advantage as, unless the ice crust is unusually thick, they sink into the snow.

Besides eating the leaves of evergreens, Snowshoe Hares also eat twigs and bark. When a hare bites off a twig, the neat, slanted surface of the cut shows that it was the work of very sharp teeth; the presence of "snowshoe" tracks, and of typical round "rabbit" scats reveal the presence of Snowshoe Hares. When they chew bark, they often kill shrubs and young trees by girdling them.

Snowshoe Hares are famous for the way their populations fluctuate. In some years they are exceedingly scarce, while in others their populations "explode." It is believed that they cause these tremendous fluctuations by their own actions. When they are abundant, they eat so much that they kill off the plants that would otherwise provide fodder for young hares; the result is a population "crash," as numbers of them starve. Then, while hares are scarce, the vegetation recovers. In time it recovers to the point at which food is available for plenty of hares again, and they undergo another population explosion. The cycle repeats itself indefinitely. When they are at the top of their abundance cycle, hares can damage forests of young trees severely. A wildlife biologist once reported finding a 10-hectare (25-acre) tract of young Jack Pines, containing more than a million trees, that was "home" to about 1000 Snowshoe Hares. Between them, they stripped off the bark and branches of all but 40 of the pines.[3] Attacks such as this are exceptional, however, and Snowshoe Hares are not usually regarded as pests.

Most of the small mammals of the forest are rodents. There are two important groups: the mouse family, which includes (with other animals) the mice and voles; and the squirrel family, which includes (again with other animals) the squirrels, flying squirrels, ground squirrels, and chipmunks. Of the many species of mice and voles, the commonest is the Deer Mouse. There are plenty of them in the forests, as well as in a variety of other habitats, and they eat the seeds of conifers as well as those of numerous other plants. But they could get along quite well even if there were no conifer forests, which is not true of the four members of the squirrel family to be considered now: the Red Squirrel, the Northern Flying Squirrel, the Yellow Pine Chipmunk, and the Golden-mantled Ground Squirrel. Nearly all the members of these four species depend on conifer seeds to get them through the winter; the only exceptions are the comparatively few Red Squirrels that spend their lives in the deciduous forests of the East.

The best known of these species is the Red Squirrel; it is familiar to everybody. We can count on seeing Red Squirrels on almost any trip into

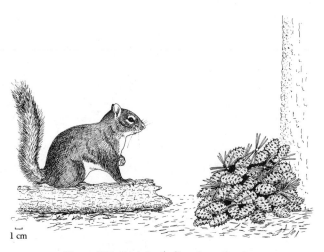

1 cm

Figure 115. Red Squirrel and cone cache.

coniferous forest; they force themselves on the attention of the most unobservant hiker with their noisy chattering, and they are always a pleasure to watch in spite of being so common. In summer, Red Squirrels have a varied diet: Buds, flowers, bark, seeds, berries, mushrooms, and insects form part of it. They also eat meat; they take nestlings from nests, and grouse chicks as they run behind the mother hen; they are also agile enough to catch mice and voles. Life is easy in summer. But squirrels are prudent. They foresee the rigors of winter and prepare accordingly, by collecting huge numbers of unopened, seed-filled cones of spruce and pine. They forage through the treetops, nipping off cone-bearing twigs, and allowing them to fall to the forest floor. Every now and again they descend, gather up the cones they have harvested, and cache them. They also collect a variety of mushrooms, dry them, and store them for winter. Newly stocked cone caches are usually concealed in such places as hollow trees or holes in the ground and are therefore seldom noticed; but occasionally a cache is piled up right out in the open, usually at the foot of a tree, like the cache of lodgepole pine cones in Figure 115 (I found this cache in the Cypress Hills of Alberta).

As winter advances and a squirrel uses up its store of cones, its cache becomes a *midden*, a hillock of discarded cone-scales and stripped cone cores left behind when the squirrel extracted the seeds. As the squirrel has no reason to conceal its garbage, middens are quite conspicuous and are far more often found than newly prepared caches. On a hike through squirrel country we are almost certain to come across them. Big ones may consist of

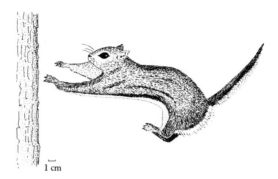

Figure 116. Northern Flying Squirrel.

1 cm

the accumulated cone remains of many years, as a squirrel will eat at the same spot year after year.

From the point of view of foresters, Red Squirrels are both pests and blessings. Undoubtedly they eat large quantities of viable tree seeds that might have germinated had they not been eaten. But they also leave a lot of seeds safely "planted" in caches that they started and then forgot about. In this way, they do a lot of reforestation. On balance, it can probably be said that Red Squirrels benefit the forests they live in.

Another squirrel with much the same diet as the Red Squirrel is the Northern Flying Squirrel; it too caches a supply of cones for the winter. Because they are nocturnal, flying squirrels are far less often seen than Red Squirrels. The time to look out for them is in the first hour or two after sunset (or the equivalent period before sunrise if you have insomnia). Neither Red Squirrels nor flying squirrels hibernate. They remain active in all but the coldest weather and hole up for protection only when the temperature falls below about −20° Celsius (−4°F). It is therefore not surprising that their survival depends on well-stocked food caches. But caches are also made by some hibernators, for example chipmunks and ground squirrels.

One species of chipmunk and one species of ground squirrel merit discussion here, since they are never found far from conifers. They are the Yellow Pine Chipmunk and the Golden-mantled Ground Squirrel. Unlike Red Squirrels and Northern Flying Squirrels, which can be found in conifer forests throughout our area, these two species are found only in the western mountains, usually in open woodlands of Douglas-fir and Ponderosa Pine (otherwise known as Yellow Pine, hence the name of the chipmunk).

The two species are rather similar in appearance (see Figure 117) and someone seeing a Golden-mantled Ground Squirrel for the first time often

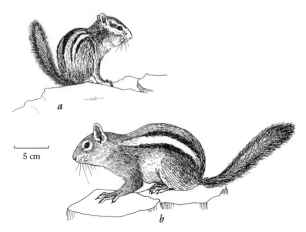

Figure 117. (*a*) Yellow Pine Chipmunk; (*b*) Golden-mantled Ground Squirrel.

mistakes it for a chipmunk. Familiarity with the two species makes the contrast between them apparent. The ground squirrel has no stripes on its face, is noticeably larger than the chipmunk, and is comparatively stolid. The Yellow Pine Chipmunk, like all other chipmunks (there are several species that all resemble one another closely) is forever dashing hither and thither energetically and seems never to need a rest; it usually holds its tail vertically upright as it runs. The Golden-mantled Ground Squirrel is not nearly so active; its movements are slower and it holds its tail at an oblique angle rather than upright.

The two animals differ also in the preparations they make for winter and the way in which they hibernate. Both of them store seeds and their caches contain an abundance of conifer seeds from the trees in their neighborhood. But unlike Red Squirrels and flying squirrels, which cache whole cones, these two species first extract the seeds from the cones, stuffing them into their cheek pouches as they do so. When they have crammed in all that the pouches will hold, they carry them to a suitable storage place in their burrows before setting off for another load. Chipmunks and squirrels with bulging cheeks are a common sight in many places in the mountains in late summer.

Although both species hibernate, the chipmunk hibernates much more lightly than the ground squirrel. A chipmunk does not fatten up before hibernating and awakes from time to time during the winter to have a meal from its cache. The ground squirrel, on the other hand, does fatten up and is able to live on its fat, without a meal, all winter. Presumably its untouched seed cache is a useful supply of easily available food when it awakes in spring to begin the chores of breeding. Both species often leave untouched caches full of seeds. Sometimes an animal appears to forget the whereabouts of a cache, sometimes it stores more than it needs, and sometimes it is killed before having a chance to use its stored supplies. As many of the "planted" seeds in abandoned caches grow into seedlings that replenish the forest, it is unlikely that a forest is damaged even when large quantities of seeds are eaten.

MORE MAMMALS: INSECT-EATERS AND MEAT-EATERS

The nonvegetarians of the northern forests belong to two radically different groups. There are insect-eaters, or insectivores, with two "families": the shrew family and the mole family. These animals are not rodents, though they are often mistaken for them. And there are true meat-eaters (insect flesh doesn't rate as meat), or carnivores, with five families: the bear family, the raccoon family, the weasel family (the fur-bearers), the cat family (cougar, lynx, and bobcat), and the dog family (wolves, coyotes, and the various species of fox). These two groups are entirely unlike each other anatomically; the insectivores are much more primitive than the carnivores in the evolutionary sense.

Mammals of all these families are to be found in coniferous forest from time to time. Here we can consider only four, one shrew, one cat, and two members of the weasel family. None of them hibernates.

Shrews are tiny animals, very similar to mice in overall appearance but easily distinguishable from them by having long, tapering snouts. They have short hair, visible ears, and little beady eyes buried in their fur. The commonest, and most wide-ranging of the several species, is the Masked Shrew, and the reason it deserves mention in a book on conifers is that it is a voracious feeder on the cocoons of the sawflies that do such damage to pines and larches (see Chapter 6). Therefore from the human point of view (also from their own and that of the trees, of course) they are most

Figure 118. Masked Shrew.   1 cm

praiseworthy animals. Sometimes, if a commercially valuable forest is severely infested with sawflies and there is a shortage of shrews locally, foresters live-trap shrews wherever a large population can be found and take them to the infested forest for release. They can do a useful job of biological control.

Shrews are seldom seen since they are exceedingly furtive. You are unlikely to find one merely by searching. It is all a matter of luck, but laziness also helps; the naturalist who sits quietly in the woods for long periods, especially if the woods are infested with sawfly, has the best chance of seeing one. Masked Shrews are those most often seen. The name is misleading; they have no conspicuous mask.

Next for the carnivores, in particular the cats: All three species in our area (cougar, or mountain lion; bobcat; and lynx) may be found in coniferous forest at times, but only the lynx is restricted to the coniferous forest; the other two species have more eclectic tastes and can be found in a variety of habitats. Everyone knows what lynx look like, but they are rarely seen, being wary, silent, and nocturnal. They eat a variety of mammals and birds, and also carrion; but their mainstay is Snowshoe Hare. Because of their dependence on hares, their numbers fluctuate wildly. The cycles of Snowshoe Hare populations were described earlier in this chapter. Parallel fluctuations affect the lynx population, of course, and for exactly the same reason.

The other two carnivores in our area that are true coniferous-forest-dwellers are the marten and the fisher, both members of the weasel family.

5 cm

Figure 119. Lynx.

Other fur-bearers of the family, such as mink, ermine, and the fearsome wolverine also live close to conifers; in the northern forests it is difficult not to. But martens and fishers are not merely forest-dwellers, they are to some extent tree-dwellers; they spend much of their time high in the trees. They are seldom seen. This is true of all carnivores, of course, and is to be expected; a carnivore must master the art of concealing itself or it will starve. The best chance of seeing one is to pay close attention whenever a group of birds such as ravens or jays is found swooping noisily at one spot. They may be harassing a marten or a fisher (even if they are not, the object of their attentions is likely to be interesting whatever it is). The chances are better of seeing the tracks of martens and fishers than the animals themselves, and a place to look out for tracks is along fallen, snow-covered logs. When they are traveling on the ground, they seem never to pass up an opportunity to run up sloping logs.[4]

Martens and fishers resemble each other in appearance and behavior, but the fisher is both much rarer and much bigger. Martens average about 60 cm (2 ft) in length, including the tail, whereas fishers are about 1 meter (3 ft) long. Because of their larger size, fishers have a greater choice of prey. Martens eat mice and voles, Red and Northern Flying squirrels, Snowshoe Hares, birds, and indeed any warm-blooded prey that isn't too big for them to tackle, as well as carrion. In summer they add insects and a variety of berries to their diet. Like the "vegetarian" Red Squirrel, which often eats meat, the "carnivorous" marten is happy to eat berries.

A fisher's diet includes all the food that martens eat plus some larger prey, including occasionally martens. They are even big enough to kill

Figure 120. Marten.   5 cm

young deer. And they are one of the few carnivores (the wolverine is another) that can succeed at the difficult task of killing and eating porcupines. The ability of fishers to dispatch porcupines has made them useful to foresters. Sometimes, when a valuable tract of young trees is ravaged by hungry porcupines, the porcupines are controlled by bringing in a population of fishers, live-trapped somewhere else, to prey upon them.

BIRDS: BROWSERS AND SEED-EATERS

The number of different bird species one may happen to come across in coniferous forest is, of course, much greater than the number of mammal species. To keep the discussion that follows to a reasonable length it has been necessary to limit consideration only to the most "deserving" species.

As with the mammals, we can to some extent classify forest birds according to what they eat. There are browsers, seed-eaters, insect-eaters, and carnivores. But birds are no more hidebound than mammals. Many of the seed-eaters eat plenty of insects as well, especially during the breeding season, and many insect-eaters eat seeds when seeds happen to be abundant and insects scarce. Berries of various kinds, when they are in season, are also an important food source. At least two of the birds to be described have such exceedingly varied diets that they deserve to be called omnivorous.

First for browsers: Two birds habitually browse on the leaves and twigs of conifers all through the winter. They are the Spruce Grouse, or Foolhen, which can be found right across the continent, and the Blue Grouse, a larger grouse of the western mountains. One usually thinks of grouse as ground birds, but in winter these two species spend much of their time

5 cm

Figure 121. Spruce Grouse.

perched high up in coniferous trees. The trees' foliage provides shelter from cold winds, in addition to easily accessible food that is not, like ground vegetation, buried under a thick layer of snow.

Coming to the seed-eaters, three species immediately recognizable as seed-eaters from the shapes of their beaks are the Pine Grosbeak, the Red Crossbill, and the White-winged Crossbill.

All three species usually build their nests in conifers. The Pine Grosbeak, like all grosbeaks, has a large, strong, conical beak, perfectly adapted for crushing seeds, and it has a greater preference for conifer seeds than the other grosbeaks. The crossbills are closely related to the grosbeaks but are smaller, only about two-thirds as long. Both groups belong to the same family as finches, sparrows, and buntings. Crossbills are instantly recognizable at close range because of their unique crossed beaks. Also at close range, the two crossbill species are easy to tell apart; only the White-winged has white wing bars, and its color is pinkish red in contrast to the duller brick red of the Red Crossbill. The Reds are most likely to be found on pine, and the White-wingeds on spruce or fir, but this is by no means an unbroken rule. The two crossbills and the Pine Grosbeak are not so easy to distinguish when seen at a distance, or in deep shade, when size is hard to judge and the details of beak structure cannot be seen. Then we can only hope that a crossbill will give itself away by its method of moving around. Like a miniature parrot, it will grip a twig with its beak while it shifts its feet, then hold on with its feet while it shifts its beak. The crossed mandibles are ideally constructed for forcing cone scales apart so that the seeds can be picked out.

Two of the most interesting birds that depend on conifer seeds for the bulk of their food are both westerners. They are Clark's Nutcracker and Steller's Jay, two handsome birds that are well known to everybody who

a

1 cm

Figure 122. (*a*) Pine Grosbeak; (*b*) White-winged Crossbill.

b

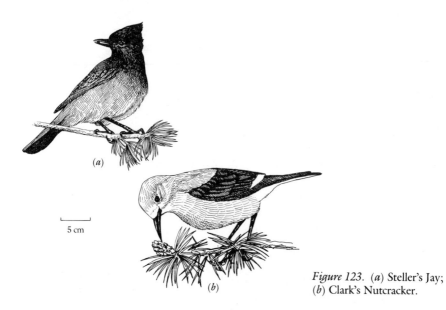

5 cm

Figure 123. (*a*) Steller's Jay;
(*b*) Clark's Nutcracker.

visits the western mountains. The nutcracker is dove-gray with conspicuously patterned black and white wings and tail; the jay is a very dark bird
with the deep blue of its body blending into a black, crested head. Their
foraging behavior has been closely studied by two ornithologists, Stephen
Vander Wall and Russell P. Balda,[5] who found that these birds cache seeds
in the same way that chipmunks and ground squirrels do, and in this way
ensure the seeding of new trees.

In our area, both birds are particularly fond of the seeds of the two stone
pines, Whitebark Pine and Limber Pine, which have large, wingless seeds.
The lack of wings is no handicap to the trees, since nutcrackers and jays do
a most effective job of sowing the seeds far from the parent trees. The
nutcracker is especially adapted to transport the seeds; indeed, nutcrackers
and pines have evolved a mutual dependence. The nutcracker has an
expandable pouch under its tongue in which it can store more than 150
stone pine seeds at one time. In late summer the birds harvest closed, seed-
filled cones from the pine trees in much the same way that red squirrels do.
They then force apart the cone-scales with their powerful, chisel-like beaks
to get the seeds out. Before tucking a seed into its pouch, a bird will inspect
it to be sure it is "edible," that is, contains a viable embryo. Thus the bird
collects only nourishing seeds; it discards the husks of empty seeds that
chanced not to be pollinated. Because of this careful checking, only seeds
capable of germinating are carried to a cache. From the trees' point of
view, only good seed is "sown." Nutcrackers may carry their seed loads for
more than 45 km (30 mi) before caching them. They like to bury their

Figure 124. Gray Jay.   5 cm

stores in open, windswept areas where snow is unlikely to lie for long, so that the seeds will be easily accessible for most of the winter. Groups of several birds make caches close together in communal caching sites. As with all cache-builders, they are apt to cache more than they need and the leftovers germinate. When, on exposed, windswept mountain slopes, we find a hectare or two where stone pines are growing quite thickly, it is probably the one-time site of a nutcrackers' communal caching ground.

Steller's Jays are also cache-makers, but jays haven't developed the art to quite the level that nutcrackers have. They have no special carrying pouch and therefore cannot carry so many seeds per trip. And instead of using communal caching grounds, each jay keeps its cache private, and drives off intruders.

Turning now from birds with specially evolved feeding habits, the next bird to mention is another jay of the coniferous forests, namely, the Gray Jay, as it is officially called. It is indeed a gray-colored jay, but it seems a shame to bestow so uninteresting a name on a bird known to nearly all campers and travelers in the woods by the colorful name of Whisky-jack (whether "whisky" or "whiskey" seems to be uncertain). The whisky-jack is, as everybody knows, a very unspecialized feeder. Technically it is an omnivore; colloquially it is a scrounger, fond of accepting any camp scraps it is offered, and of helping itself to any it is not offered. In uninhabited regions it probably divides its attention between seeds and insects.

More Birds: Insect-eaters and Meat-eaters

Two groups of insect-eating birds familiar to everybody are the chickadees and the nuthatches. They eat seeds as well, but insects form the greater part of their diets. They often forage in mixed flocks and are often to be found in coniferous or deciduous forests anywhere in our area. The chickadee species most closely associated with conifers is the Mountain Chickadee of the western mountains. It is easy to recognize even when, as often happens, it forages in a mixed flock with other species of chickadee, since it is the only chickadee with a white eyebrow line. Two species of nuthatches are conifer specialists; they are the Red-breasted Nuthatch, which occurs from east to west across the continent, and the Pygmy Nuthatch of the Ponderosa Pine forests of the western interior.

Chickadees and nuthatches are voracious insect-eaters and we can observe their different foraging techniques by watching a mixed flock doing a minutely thorough "clean-up" through the crown of a tree. Insects of certain kinds are as abundant in winter as in summer, but in winter they merely survive passively in a dormant state. Depending on the species, some insects overwinter as adults, others as eggs, larvae or pupae; there are literally thousands of them on any tree, tucked away in myriad cracks, crannies, and crevices, awaiting renewed activity when spring arrives or else swift death as the quarry of a hungry bird. Insects tucked away in the bark of a tree's trunk or branches run the risk of being found by nuthatches, which creep up and down and around them searching every crack. Chickadees, instead of creeping along thick, solid branches, do their foraging among the flexible tips of the ultimate twigs. No matter how much a twig sags and swings, a chickadee can cling to it effortlessly, often hanging upside down, until every concealed insect has been found and eaten.

1 cm

Figure 125. (*a*) Red-breasted Nuthatch; (*b*) Mountain Chickadee.

Nuthatches and chickadees cannot get at insects protected by a layer of unbroken bark. But these insects are accessible to woodpeckers, which destroy large numbers of such pests as bark beetles (*Dendroctonus* species), engraver beetles, and the grubs of the wood-boring flathead borers (see Chapter 6). Woodpeckers of many species can be found on coniferous trees at any season, noisily hammering at tree trunks and branches in pursuit of their prey. Most woodpeckers are as likely to feed on deciduous as on coniferous trees, but the two species of three-toed woodpecker (the Northern Three-toed and the Black-backed Three-toed) prefer conifers. They are especially likely to be found foraging on standing dead trees that have been killed by beetles or else on fire-killed trees, which also harbor numerous wood-eating insects.

Most of the birds mentioned above can be found in our area all through the year. Pine Grosbeaks and Red-breasted Nuthatches may shift their activities southward for a short distance in winter, but even so all these birds are well able to fend for themselves in the woods at all seasons. There are other insect-eaters that seem less well adapted to the cold, though they still rely on dormant insects in the trees for winter fodder. These are the wrens and kinglets; some of them stay around, but only in the south-ernmost part of our area, all through the year. One of the wrens and both the kinglets are especially fond of conifers. The wren that inhabits conifer-ous forest is the Winter Wren; its French name is *le Troglodyte des forêts*. It draws attention to itself with its song—a loud, prolonged series of short tinkling notes, varying rapidly between high and low, and broken by occasional short pauses. Even when this distinctive song sounds close at

Figure 126. Northern Three-toed Woodpecker.   5 cm

Figure 127. (*a*) Pine Warbler; (*b*) Golden-crowned Kinglet; (*c*) Winter Wren.

hand it is often impossible to see the singer. It is a tiny, brownish bird and is likely to be concealed deep in a dense, dark thicket of impenetrable undergrowth near the ground. It spends most of its time close to the ground, hunting for insects. This explains why it must spend the winter where the snow frequently melts; the deep, long-lasting snowbeds of high latitudes would cover its food supply.

The Golden-crowned and Ruby-crowned kinglets are also coniferous forest dwellers, and they also shift their range southward for at least a short distance during the winter, though not for the same reason as the Winter Wren, as both kinglets do their foraging in the tops of the trees. One other small insect-eater deserves mention here: the Pine Warbler, which is the only wood warbler closely associated with conifers, specifically pines. In our area it occurs only in the central part of the continent, west of the Appalachians and east of the prairies, and there only in summer, where it creeps over the branches and trunks of pine trees in an incessant search for insects.

The thrushes are another group of forest birds often to be found among conifers, but only one thrush species is seldom found anywhere else: the Varied Thrush of the West. Its favorite habitat is full-grown conifer forest at its dampest and darkest. It closely resembles an ordinary American

Figure 128. Varied Thrush.   1 cm

Robin, except that it has an orange-red eyebrow line and wingbars as well as an orange-red breast, and the breast is crossed by a broad black band. It is so like a robin in general appearance and mannerisms that easterners seeing one for the first time sometimes assume that what they are seeing is an aberrant robin rather than a bird of a different species. The Varied Thrush bridges the gap between insect-eaters and meat-eaters. Besides eating true insects such as ants and ground beetles, and "insect-textured" prey like woodlice, centipedes, and millipedes, it also enjoys the meat of slugs and snails, which are plentiful in its damp habitat. The bird can be called omnivorous as it eats berries and seeds as well. It is a ground-feeder, and spends most of its time in dense undergrowth near ground level.

Now for the true meat-eaters, the hawks and owls: None of the diurnal birds of prey—that is, eagles and hawks in the general sense, including the falcons, accipiters, and buteos—are habitual dwellers in coniferous forest. The species most likely to be seen in such a setting is the Goshawk. But the nocturnal predators, the owls, are different. Many owl species live and

Figure 129. Boreal Owl.   5 cm

hunt in the evergreen forests, for instance, the Boreal Owl (see Figure 129), the Great Horned Owl, the Hawk Owl, the Great Gray Owl, the Barred Owl, the Long-eared Owl, the Saw-whet Owl, and (only in the West) the Pygmy Owl. All of them eat large quantities of mice and voles. Owls are usually more numerous than they seem.

Like all hunters, their survival depends on stealth, and the naturalist who doesn't deliberately look out for them is unlikely to see them, especially as most of their activity takes place in the dark of night. But anybody hiking or skiing through the forest (winter is an especially good time because there are fewer distractions) may be rewarded for keeping a careful lookout by seeing an owl silently and inconspicuously roosting on a branch of an evergreen.

NOTES

1. Jamie Swift, *Cut and Run* (Toronto: Between the Lines Press, 1983).

2. Island Protection Society, *Islands at the Edge* (Seattle: University of Washington Press, 1984).

3. W. Rowan, "Reflections on the Biology of Animal Cycles," *in Journal of Wildlife Management*, vol. 18, pp. 52–60, 1954.

4. L. L. Rue, III, *Furbearing Animals of North America* (New York: Crown Publishers, Inc., 1981).

5. Stephen Vander Wall and R. P. Balda, "Remembrance of Seeds Stashed," *in Natural History*, vol. 92, no. 9, pp. 61–64, September 1983.

Chapter 10

Companions of Conifers

Preceding chapters have discussed a number of the plants and animals that live in the evergreen forests, and the ways in which they affect, and are affected by, the coniferous trees. This chapter deals with the largest of the conifers' fellow organisms: hardwood trees. To keep the chapter within bounds, we shall consider only the hardwoods that could reasonably be called "tenants in conifer country" or "companions of the conifers"; this description does not include the numerous hardwoods of the eastern deciduous forests—maples, oaks, beeches, elms, ashes, and so on—where such conifers as white pines and eastern hemlocks are companions of the hardwoods rather than the other way around.

Four Hardwood Genera

The hardwood companions of conifers are not numerous. There are only four large hardwood trees that play an important part in the life of the evergreen forests. They are Balsam Poplar (*Populus balsamifera*); Trembling Aspen, also called Quaking Aspen (*Populus tremuloides*); Paper Birch (*Betula papyrifera*); and Red Alder (*Alnus rubra*). And besides these large trees, there are a number of different species of willows and alders, most of which are shrubs, although some grow to tree size if conditions are unusually favorable.

The hardwoods that concern us, therefore, belong to four genera: the poplars (*Populus*), the willows (*Salix*), the birches (*Betula*), and the alders (*Alnus*), and in some respects they resemble conifers. All of them bear

153

catkins. A catkin consists of densely packed rows of flowers, growing along a short stalk that may be upright or dangling. The individual flowers are tiny and devoid of petals. There are female catkins, made up of female, seed-bearing flowers, and male catkins, made up of male, pollen-producing flowers. The noteworthy point is that some catkins look remarkably like small conifer cones. The resemblance is especially close in the case of the female catkins of alders; they are so conelike to look at that they are often called "cones." And many male catkins are similar in appearance to conifers' pollen cones. Figure 130 shows representative female and male catkins from each of the four genera.

The resemblance of catkins to cones is not surprising when we consider their purpose. Both are the reproductive organs of plants that live in high latitudes, where wind pollination is more likely to succeed than insect pollination. The female flowers (in hardwoods) and the seed cones (in conifers) must be pollinated early in the year if their seeds are to ripen in the short growing season, and there is always a risk that the weather will be

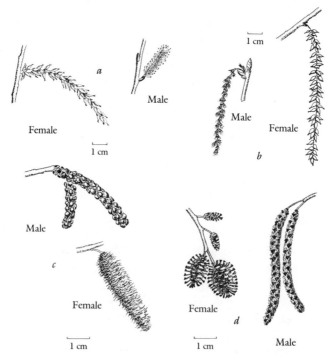

Figure 130. (*a*) Willow catkins; (*b*) poplar catkins; (*c*) birch catkins; (*d*) alder catkins.

too cool for insects (especially bees) to be on the wing when the trees are flowering. Thus catkins, like cones, are adapted to wind pollination. The poplars, birches, and alders are wholly dependent on the wind to pollinate them, but the willows are wind-pollinated only in cool springs. If the weather happens to be warm and insects active when the willows are in flower, then they are insect-pollinated. On a warm spring day, pollen-covered pussy willows (which are the male catkins of some species of willows) attract numerous bees.

The four catkin-bearing genera belong to two families. The willows and poplars belong to the willow family, Salicaceae, in which each plant (tree or shrub) bears either male or female catkins (not both) and in which the tiny seeds bear long, downy hairs (the "cotton" of cottonwoods) so that even a gentle wind will carry them far from their parents. The birches and alders belong to the birch family, Betulaceae, in which each plant bears catkins of both sexes, and in which the seeds are winged. Thus all the catkin-bearing plants, like the majority of conifers, have seeds carried by the wind.

Of the four groups of "conifer companions" (willows, poplars, birches, and alders) the willows are probably the least important from the point of view of their effects on conifers. They are companions only in the sense that they grow, in suitable sites, throughout the north country. The great majority of willows are shrubs rather than full-sized trees, and it is rare for large numbers of them to occupy ground where conifers might otherwise live. Many species grow only where water is close by, along the shores of lakes, rivers, and streams; others are dwarf shrubs of the tundra, either above the tree line on mountains, or north of the tree line in subarctic latitudes.

The majority of large willows have long, narrow leaves. In the dwarf willows of the tundra, although many species have short leaves, one can usually find enough specimens to discover whether the two recognizably different kinds of catkins (males and females) grow on separate plants. This is a sure sign that the plants are willows, since the only other catkin-bearers with the sexes separate are the poplars and there are no dwarf poplars. The willow genus as a whole is therefore not hard to recognize. But distinguishing the separate species is another matter; there are about 75 of them in our area, not counting numerous hybrids, and the majority are exceedingly difficult to identify with certainty. No more need be said about willows. Here we concentrate on the hardwoods that do have an appreciable effect on the welfare of the conifers. Fortunately, they are easy to identify.

POPLARS

The poplars in general (that is, all the trees known as poplars, cotton-woods, and aspens) are fast-growing, hardy trees. Life is usually rather short for each individual tree, but they propagate so rapidly that there is little risk of their numbers decreasing as long as the environment they need does not diminish. Two poplar species, Balsam Poplar and Trembling Aspen, are adapted to withstand the intensely cold winters of northern Canada and Alaska (see Chapter 1), and can therefore grow as far north as the tree line in the company of the similarly adapted White and Black Spruce, Balsam Fir, Jack Pine, and Tamarack. They are found, either as tracts of pure poplar forest, or else mingled with the evergreens, almost everywhere evergreen forests grow.

The two species are easy to recognize. Trembling Aspen is famous for its trembling leaves; the reason they tremble, even in the gentlest breeze, is that they have flat petioles (leafstalks). If you look at an aspen leaf lying with its blade flat on your hand, you see the strap-shaped petiole edge-on. The leafblade itself is almost circular. Trembling Aspens also have distinctive smooth, powdery, greenish-white bark. If you rub the bark, it whitens the skin of your hand much as a piece of chalk would. The bark does not loose its characteristic smoothness until the tree is nearly at the end of its short life-span. Trembling Aspens seldom live for more than 100 years before succumbing to diseases that cause them to decay and die.

Balsam Poplar has larger, more tapered leaves, with unflattened petioles; therefore the leaves don't tremble. It has deeply furrowed bark. And it is deservedly famous for the wonderful fragrance of its big, sticky buds in spring. Like Trembling Aspen, it is a short-lived tree. Many westerners are

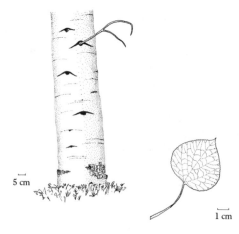

5 cm

1 cm   *Figure 131.* Trembling Aspen.

familiar with an almost identical western tree that used to be known as Black Cottonwood. It often grows to tremendous size in the luxuriant forests of the West Coast. Balsam Poplar and Black Cottonwood hybridize so easily, however, that nowadays the two trees are regarded as merely geographic races of a single species, which has been given the name Balsam Poplar.

Trembling Aspen and Balsam Poplar resemble each other in that neither can live with their crowns shaded. If any kind of poplar is to live, its leaves must receive full daylight. They also resemble each other in having shallow, but very wide-spreading root systems; the importance of this to their welfare will become clear below. The two species require rather different environments. Balsam Poplar needs a moist soil, and therefore grows best on low-lying ground, especially valley bottoms. Trembling Aspen can get by with much less water, and does fairly well even on the driest hillsides. Between them, the two species can take possession of almost any tract of country *provided* their seeds arrive there at an appropriate time.

This is the crux of the problem for all species of poplars. They produce enormous quantities of seeds, but the seeds will not germinate unless conditions are suitable precisely at the time they reach maturity and fall from the parent tree. The reason is that they remain alive for only 2 or 3 weeks after falling. Poplars are most unusual in this respect. In the great majority of plants the seeds remain viable (capable of germination) for at least a year after they mature, usually for several years. But for poplar seeds to germinate and begin growth, they must have plenty of light and moisture without delay; they will not wait. As they cannot endure shading by

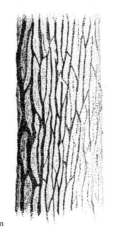

5 cm

1 cm

Figure 132. Balsam Poplar.

other plants, they succeed only if the wind deposits them on bare, or nearly bare, ground. And because of their intolerance of shade, they run the risk of being dried out by sunshine and dry air in a shadeless environment.

The circumstances in which a new forest of poplars can get started are therefore uncommon. The open, unoccupied ground that they need is most often the result of a forest fire, perhaps in a spruce or jack pine forest. The fire must happen in a place where fresh poplar seeds are available to seed the newly burned area; it must happen at a season of the year when no competing plants have had a chance to establish themselves before the poplar seeds arrive. And there must be enough water available in the very short period that the seeds are viable for them to germinate and for the young seedlings to get started. The water can come as rain or, for balsam poplars, from the waterlogged ground of river floodplains and sandbars. Only on the rare occasions when all these special conditions are met will poplars invade a new area successfully. The opportunity comes much more rarely for Trembling Aspens, which must rely on well-timed rain for their water supply, than for Balsam Poplars growing in moist, valley-bottom soil.

Once poplars have invaded, they are tenacious. The reason is that they do not spread only by seed; they also send up new shoots (root suckers) from their wide-spreading roots. Trembling Aspens, especially, depend on root suckers to perpetuate themselves because their seeds so seldom germinate. A single "parent" tree may have a root system 50 m (80 ft) across, and this whole large area soon becomes populated with root suckers that quickly grow up into new trees. The word "parent" is in quotation marks because this method of propagation is not reproduction in the ordinary sense; no genetic mixing is involved. Genetically speaking, the new trees are merely parts of the "parent" tree, not its offspring. After a seedling aspen has succeeded in establishing itself (a rare event, as we have seen), it will, in time, produce large numbers of new trees from root suckers. These, in turn, will produce their own root suckers, and so on indefinitely. Only a catastrophe such as disease or fire will halt the process. In this way extensive clones of Trembling Aspens are formed, each many hectares in area. The trees forming a single clone (there may be several hundreds of them) are genetically identical.

The fact that a forest of Trembling Aspen is a patchwork of different clones is often easy to see, especially in spring and fall. In spring, all the trees in a clone (which are all of the same sex, of course) come into flower and then into leaf at the same time as one another, but at a different time from the trees in neighboring clones. A forest of Trembling Aspen in spring is therefore a mosaic of slightly different greens and gray-greens, each patch of the mosaic (each clone) displaying the color corresponding to its own stage of flowering or leaf-unfolding. The pattern becomes visible again in the fall. The leaves of Trembling Aspen turn from green to brilliant gold before they are shed. As in spring, the trees within one clone are synchronized, but the separate clones are not. As a result the clones form a mosaic of distinct patches, each of a different shade of green or gold.

Now we come to the effects of poplars on conifers, and of conifers on poplars. The seedlings of shade-tolerant conifers, especially spruces, benefit during their early years of growth by having a "cover" of poplars to protect them. A forest in which young evergreens are growing up among taller, overshadowing poplars is a common sight in the north country. The poplars serve as "nurse" trees for the conifer seedlings. In spring and summer they provide a damp, shady environment for the seedlings; in the fall they supply a mulch of leaf mold that tends to conserve soil moisture. And in winter the poplar trunks act as snow fences, causing snow drifts to accumulate among them; the deep snow engulfs the young conifers and insulates them from the cold. As a final benefit, the poplars die young; they do not compete for soil nutrients with the rapidly growing conifers once their usefulness as nurse trees has ended. This happy mix of trees does not occur in every poplar forest, of course. For a poplar forest to acquire conifer "wards," conifer seeds must be blown in and settle among the poplars.

Although poplars benefit conifers, they receive no reciprocal benefit. Poplars and evergreens compete with each other and the poplars would steadily lose ground to shade-tolerant conifers if they did not, occasionally, have a chance to claim new territory and so make up for what they have lost. Such a chance comes only when a conifer forest burns to the ground at a time when poplar seeds are available and conditions are right for them to germinate and become established. Poplars, like Jack Pines and Lodgepole Pines (see Chapter 8) depend on fires for their survival. When a tract of upland forest burns, the new environment that the fire creates is often equally suitable for Trembling Aspen or Jack Pine (Lodgepole Pine is the alternative in the West) and whether the first post-fire forest is made up of pines or aspens may simply depend on which seeds get there first.

Before leaving the poplars, one other *Populus* species deserves mention. It is Bigtooth Aspen (*Populus grandidentata*), which is very like Trembling Aspen except that its leaves are almost twice as large and have toothed margins. Its geographic range is not nearly as great as that of Trembling Aspen; the northern limit of its range probably coincides with the line on the map north of which minimum winter temperatures fall below $-40°$

1 cm

Figure 133. Bigtooth Aspen.

Celsius (see Chapter 1). It cannot endure, as Trembling Aspen can, the intense cold of the Far North. Moreover, it is an eastern tree, occurring no farther west than Minnesota and eastern Manitoba. Within its limited range its ecology is similar to Trembling Aspen's; the two aspens often grow together as conifer companions in eastern forests.

BIRCHES

Of the few hardwoods to be found in evergreen forests, Paper Birches are the most familiar and to many they are also the most beautiful. They act as nurse trees for young evergreens because, like Trembling Aspens, they are intolerant of shade, fast-growing, and short-lived; they rarely survive for more than 90 or 100 years. And like Trembling Aspens they are well adapted to thrive in forests that are prone to frequent forest fires. Anyone who has used shreds of bark from a Paper Birch for kindling knows how well it burns, even when wet. This certainly suggests (though it doesn't conclusively prove) that, far from being a disadvantage to the species, the thin, flammable bark is probably an advantage. Paper Birch is better adapted than Trembling Aspen in this respect. It is true that the bark cannot protect an individual birch tree from being killed by a fire, but it certainly fuels the fires that destroy the evergreens in whose shade neither birch nor aspen can live.

Both species are therefore "fire-dependent" (see Chapter 8) and can hold their own only in a forest that is regularly burned or, if the forest is exploited, regularly cut over. They must, at all times, have plenty of light; anything less than full daylight is inadequate. Not merely the tree as a whole, but each leafy branch requires light. This is why full-grown aspens

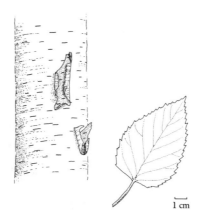

Figure 134. Paper Birch.   5 cm   1 cm

and birches growing in closed forests have tall, straight trunks, free of branches. The low branches they had in youth, when they and the trees around them were saplings, have long since atrophied from lack of light and dropped off.

Paper Birch differs from the poplars in that it does not produce root suckers. It relies on seeds to spread even a short distance. New trunks often sprout from the base of a cut or burned stump, however, and produce a many-stemmed tree to replace the one that was destroyed. The development of a many-stemmed birch is different from that of a forked or many-stemmed evergreen. New trunks on an evergreen can develop only when existing side branches bend and grow upward, which they do if the terminal bud of the original stem is killed (see Chapter 9). The new shoots developing on a birch stump are not preexisting side branches, they have grown from scratch from the tissues of the stump.

Paper Birch is not the only birch species to grow in the evergreen forests, though it is certainly the most distinctive. Water Birch (*Betula occidentalis*) is a small tree or shrub, common from Manitoba westward, most often found growing along stream banks with willows and alders.

Figure 135. Wire Birch.   1 cm

Wire Birch (also called Gray Birch, *Betula populifolia*) is an exceedingly common little tree in New England and the Maritime Provinces of Canada. Ecologically, Wire Birches behave like Paper Birches; they are much smaller and even shorter-lived, but do useful service as nurse trees for young conifers starting life in their shade. Their distinctively shaped leaves make them easy to recognize.

ALDERS

The alders (genus *Alnus*) are the fourth group of hardwoods that (with willows, poplars, and birches) share the northern forests with the evergreens. The most widespread alders are shrubs. Two species are common almost everywhere east of the continental divide—Speckled Alder (*Alnus rugosa*) and Green Alder (*A.crispa*)—and two other species almost everywhere west of it—Mountain Alder (*A. tenuifolia*) and Sitka Alder (*A. sinuata*). Alders are noticeable in the fall by virtue of the fact that their leaves do not turn yellow, orange, and gold like the leaves of aspens, poplars, and birches, but remain a dingy green for as long as they remain on the plants.

Besides these wide-ranging shrubs, there is also an alder that grows to be a large tree. This is Red Alder (*A. rubra*); it is found only in the West Coast forest, where it is an important nurse tree for conifers. Like the other nurse trees it is intolerant of shade and can colonize only areas that are newly burned over or cut over, where there is an opening in the deep shade of the evergreen forest. Like them, also, it is fast-growing and short-lived. But in one respect it is a far more valuable "nurse" than any poplar or birch. This is because it is a nitrogen-fixer.

5 cm

1 cm

Figure 136. Red Alder.

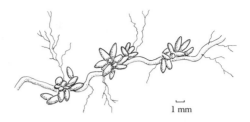

Figure 137. Nodules on Red Alder roots.

1 mm

All plants require nitrogen; they cannot live without it. They cannot, however, absorb and use gaseous nitrogen, which constitutes nearly 80 percent of the air, but must obtain it through their roots, in the form of nitrogen-containing compounds dissolved in the water in the soil. The question arises: What is the process that continually renews the soil's quota of nitrogen compounds, and ensures that the supply will never be used up? The answer is: certain species of bacteria that *can* absorb gaseous nitrogen and convert it into soluble compounds. The most abundant of these nitrogen-fixing bacteria always live in soil surrounding the roots of certain plants. The bacteria are stimulated to multiply by secretions from the plants' roots, whereupon they enter the root cells, causing some of them to swell up into nodules. Not all plants interact with nitrogen-fixing bacteria to form nodules; that is, not all plants are nitrogen-fixers. Those most familiar to farmers and gardeners are legumes such as peas, beans, clover, and alfalfa, which are often grown, not merely for their own sake, but also to enrich the soil with nitrogen for the benefit of subsequent crops to be grown on the same ground.

The nitrogen fixers of greatest importance in evergreen forests are the alders. Their roots, like those of legumes, become infected with nitrogen-fixing bacteria, which occupy clusters of orange nodules on the alder roots. It is easy to find the nodules by digging up the shallow roots of an alder. Alders are valued for their ability to enrich the soil with nitrogen compounds that other plants can absorb. This is especially true in the Douglas-fir forests of the West Coast, where the soil tends to be deficient in nitrogen. If a generation of red alders is allowed to grow to maturity on cutover land, the Douglas-firs that grow on the same land subsequently benefit greatly from the increased soil fertility.

Epilog

Evergreen forests mean different things to different people. To some they are storehouses of valuable forest products waiting to be profitably exploited. To others (presumably most of the readers of this book) they are a vital part of the biosphere. They recycle the air we breathe, ensuring that our oxygen supply will never be exhausted; they store rainwater that would otherwise be quickly lost to the land; they create and protect soil; they provide habitat for innumerable plants and animals; and they give enormous aesthetic and recreational enjoyment to millions of people.

It is easy to persuade ourselves that these two aspects of the forest are not in conflict; indeed, they need not be. But there is no question that, at present, they are. Problems have arisen because of the widely held belief that "forests are renewable." This phrase originated with foresters and has been endlessly parroted, especially by politicians, with disastrous results. The truth is that *exploited forests will not and cannot renew themselves*, given modern methods of forest exploitation.

Consider, for example, the consequences of clear-cutting a forested mountainside in the West Coast rain forest. All the plants, not only the trees, are destroyed. Indeed the whole environment is destroyed. Before cutting there was a moist, windless, deeply shaded habitat occupied by a diverse community of plants and animals: shrubs, herbs, ferns, mosses, lichens, fungi, mammals, birds, amphibians (frogs, toads, and salamanders), insects, mollusks (snails and slugs), and an uncountable array of soil organisms. There were ponds and streams with their own, equally diverse aquatic communities; the ground was everywhere covered with a blanket of spongy soil.

After the forest has been clear-cut, the habitat needed by this multitude of living things no longer exists. All the vegetation is gone; all the animals are gone; the huge volume of rainwater that was previously absorbed and transpired by the trees now flows away as torrential runoff. The soil has been crushed by heavy machinery and is washed away; landslides and mudflows surge down the mountain slopes into the valleys. What were clear, shaded, gently flowing rivers become raging torrents of turbid water full of mud and logging debris; all the aquatic life, including all the fish, immediately die. If there are high altitude "islands" of forest left surrounded by the clear-cut, they become prisons for animals such as deer, which cannot migrate to their winter habitat across open, slash-covered land. Lastly, logging causes severe ecological damage far away from the felled forests, in the lakes and sheltered seaways where log booms are stored. Fragments of bark and wood raining down from a floating log boom form a layer on the bottom that both suffocates and poisons lake-botttom and sea-bottom life; all the other aquatic organisms that depend on this life for food are inevitably affected too.

It should now be clear that to expect the forests to renew themselves is fatuous and dangerous. Wood is a renewable resource. Forests, in their original, pristine form, are not. Land from which the soil has been washed away will not bear a "complete" forest again for a very long time, perhaps for as much as 30 or 40 human generations, though it will probably support a stand of trees much sooner. However, a stand of trees is not a forest. The development of a forest ecosystem to its fullest extent, with its myriad coexisting species of plants and animals, is an extremely slow process, so slow that no exact replica of an existing forest will ever appear. Wilderness is shrinking, plant and animal species are going extinct, and the climate is continuously changing. All these things are proceeding too fast for a modern forest to renew itself in its entirety before the conditions needed to maintain it have disappeared forever.

It is sometimes argued that forests recover from clear-cutting in the same way that they recover from fire. This is not true. The average fire is far smaller in area than the average clear-cut, and the burned tract is far less severely damaged. Except in ground fires (see Chapter 8), which are seldom big, only the topmost layer of soil suffers appreciable damage; the root systems of most trees, shrubs, and herbs (though not, of course, the evergreen trees) remain capable of sprouting. The store of buried seeds is unaffected. The only elements "consumed" by a fire—in the sense that they are converted to vapor and dissipated—are carbon, oxygen, and hydrogen, and these are easily replaced. The other chemical elements in the burned trees—the mineral nutrients—remain in the ash and are immediately available to the plants that promptly grow up on a burned area. The minerals do not remain on clear-cut land, of course; they are part of the wood that is taken away.

To put the matter in a nutshell, forests are adapted to fire but not to clear-cutting. In fact, as described in Chapter 8, fires are a necessary part of most forests' life cycles; the same can hardly be said of clear-cutting! The replacement of trees on cutover land takes place most rapidly in warm, dry regions, where the fire cycle is short. It takes place much more slowly in wet or cold regions, where the fire cycle is long. But to repeat a point that cannot be too forcefully emphasized, the regrowth of trees is not the same thing as the renewal of a forest. A forest is much, much more than a collection of trees. Truly pristine forests are now so uncommon in settled parts of this continent that most people do not realize the degree to which second-growth forests, attractive though they often are, fall short of true wilderness forest.

This brings us back to the two irreconcilable demands of dwellers in forested country. We want to consume forest products, and we want to conserve the forest. The metaphor about wanting to have your cake and eat it describes the situation exactly. How can it be done?

There are only two solutions. The first is probably impracticable. The second is feasible but costly.

The first is to devise methods of harvesting trees in a way that does not destroy the forest. This would mean harvesting individual trees or, at most, small groups of trees, here and there, without using heavy, ground-crushing machinery and without making more than small, widely scattered gaps in the canopy. In a word, it would mean a return to old-style logging. It is frightening to realize that so simple a solution seems out of the question, because it would slow the rate of production of wood and wood pulp far below the rate of demand of our huge urban populations. We are victims as much as beneficiaries of technological advance. On the one hand modern technology enables us to destroy our environment; on the other it has enormously increased our demands for the so-called resources that a healthy environment can provide.

The second solution to the exploiters versus conservers dilemma is to increase the productivity of *part* of our forested lands to the absolute maximum, whatever the cost. This part, consisting of blocks of land that might conveniently be called "resource forests," should be "farmed" so diligently that it can provide all the forest products we need. Pests, diseases, and fire should be rigorously controlled, with no restriction on the methods used beyond an insistance that they should not cause contamination of any kind beyond the borders of the resource forests where they are used. New trees should be planted, fertilized, and tended to ensure complete and rapid replacement of those that are cut. The renewed forest would not and could not be the same as pristine forest, however.

All the forests that are not farmed as resource forests can then be set aside as "wilderness forests" to meet a need that is at least as great as that for forest products, the need for wild, undisturbed, uninhabited land, where complete ecological communities can maintain themselves. Wilderness forests should be given permanent protection. They should be treated as ecological reserves to preserve biological and genetic diversity. They should be available for nondestructive enjoyment. And they should be out of bounds to all forms of commercial exploitation in perpetuity. As the world becomes more crowded, the demand for wilderness forests—indeed, for wilderness of all kinds—steadily increases. There is no longer any need, if there ever was, for conservationists and environmentalists to be on the defensive; they should not be apologetic in pressing their demands.

But we shall all have to share the costs. They may be high, in terms of both money spent and luxuries foregone. As the old Spanish proverb puts it: "Take what you want," said God, "take it, and pay for it." It should, by now, be apparent to all of us that there are no free trees.

Suggested Further Reading

Banfield, A. W. F. *The Mammals of Canada*. Toronto: University of Toronto Press, 1981.

Borror, D. J., and R. E. White. *A Field Guide to the Insects of America North of Mexico*. Boston: Houghton-Mifflin Co., 1974.

Bull, John. *The Audubon Society Field Guide to North American Birds, Eastern Region*. New York: Knopf, 1977.

Brockman, C. F. *Trees of North America*. New York: Golden Press, 1968.

Burt, W. H., and R. P. Grossenheider. *A Field Guide to the Mammals*, 2d ed. Boston: Houghton-Mifflin Co., 1976.

Godfrey, W. E. *The Birds of Canada*, rev. ed. Chicago: University of Chicago Press, 1986.

Hosie, R. C. *Native Trees of Canada*, 7th ed. Ottawa: Canadian Forest Service, 1969.

Klots, A. B. *A Field Guide to the Butterflies of North America, East of the Great Plains*. Boston: Houghton-Mifflin Co., 1977.

Lincoff, G. H. *The Audubon Society Field Guide to North American Mushrooms*. New York: Knopf, 1981.

Milne, Lorus, and Margery. *The Audubon Society Field Guide to North American Insects and Spiders*. New York: Knopf, 1980.

Peattie, D. C. *A Natural History of Trees of Eastern and Central North America*, 2d ed. Boston: Houghton-Mifflin Co., 1966.

———. *A Natural History of Western Trees*. Boston: Houghton-Mifflin Co., 1953.

Peterson, R. T. *A Field Guide to the Birds*, 4th ed. Boston: Houghton-Mifflin Co., 1980.

———. *A Field Guide to Western Birds*, 2d ed. Boston: Houghton-Mifflin Co., 1961.

Petrides, G. A. *A Field Guide to Trees and Shrubs*. Boston: Houghton-Mifflin Co., 1973.

Potvin, A. *A Panorama of Canadian Forests*. Ottawa: Canadian Forestry Service, 1975.

Pyle, R. M. *The Audubon Society Field Guide to North American Butterflies.* New York: Knopf, 1981.

Rue, L. L., III. *Furbearing Animals of North America.* New York: Crown Publishers Inc., 1981.

Smith, A. H. *The Mushroom Hunter's Field Guide.* Ann Arbor: University of Michigan Press, 1973.

Swan, L. A. and C. S. Papp. *The Common Insects of North America.* New York: Harper & Row, 1972.

Swift, Jamie. *Cut and Run: The Assault on Canada's Forests.* Toronto: Between the Lines Press, 1983.

Udvardy, M. D. F. *The Audubon Society Field Guide to North American Birds, Western Region.* New York: Knopf, 1977.

Index